THE EARTH IS NOT FOR SALE

A Path Out of Fossil Capitalism to the Other World That is Still Possible

THE EARTH IS NOT FOR SALE

A Path Out of Fossil Capitalism to the Other World That is Still Possible

Peter Schwartzman
Knox College, USA

David Schwartzman
Howard University, USA

World Scientific

NEW JERSEY • LONDON • SINGAPORE • BEIJING • SHANGHAI • HONG KONG • TAIPEI • CHENNAI • TOKYO

Published by

World Scientific Publishing Co. Pte. Ltd.
5 Toh Tuck Link, Singapore 596224
USA office: 27 Warren Street, Suite 401-402, Hackensack, NJ 07601
UK office: 57 Shelton Street, Covent Garden, London WC2H 9HE

British Library Cataloguing-in-Publication Data
A catalogue record for this book is available from the British Library.

THE EARTH IS NOT FOR SALE
A Path Out of Fossil Capitalism to the Other World That is Still Possible

ISBN 978-981-3234-24-6
ISBN 978-981-3276-64-2 (pbk)

For any available supplementary material, please visit
https://www.worldscientific.com/worldscibooks/10.1142/10827#t=suppl

Desk Editor: Amanda Yun

Typeset by Stallion Press
Email: enquiries@stallionpress.com

Printed in Singapore

Foreword

It should be common sense by now that we face enormous, life-threatening challenges, of planetary proportions, thanks to a couple of centuries of reckless capitalist practices and policies, and to inadequate, if not complicit, alternatives that came out of now largely gone and discredited state-socialist systems. As Schwartzman and Schwartzman show in this book, the causes, impacts, and challenges are hardly spread evenly. Those with control over capital and of the main means of violence have been (and are) least impacted by global warming, biodiversity decline, air, water, and soil pollution, soil and ozone-layer destruction, freshwater scarcities, and other interlinked forms of ecological devastation. It is most of the rest of humanity that has been and continues to be subjected to these deadly horrors, horrors imposed on majorities by means of putting capital accumulation ahead of the majority of people and of the rest of nature. Centuries of destructive impacts by at first a handful of capitalist societies (politically dominated by a small minority) have by now spread to cover, directly or indirectly, the entirety of the planet's surface. It is historically unprecedented that a small proportion of a single species is responsible for devastation of such magnitude. And it is, in terms of percentages, really a tiny number of humans, with ever more concentrated power in their hands (via privatization — down to molecular levels, with gene patenting —, redirection of social wealth to military spending, centralization of finance and diffusion of debt, cuts in social provisions, direct and indirect subsidies to businesses, etc.), who are directly responsible for the current situation, and yet virtually unaccountable for their actions, currently or historically. Identifying individual culprits or groups, though, is of little use. Capitalists are personifications of general tendencies in societies at large. What must be overcome is a set of social institutions (or, better, the social relations that make them happen) that encourage such behavior — that make it seem normal — and that create the conditions for a small minority to undermine ecosystems with impunity and thereby wreck the lives of millions and that of their progenies.

Most technical experts (including the scientific mainstream), politicians, and activists continue undeterred in proffering capitalism-friendly ideas and injecting policies with notions and actions that fail to address or intentionally distract attention away from capitalist causes. This is most evident in the various COP, Davos, and G-7 meetings, where the most powerful and mostly unelected (and largely upper-class and male) meet to discuss what they deem to be topmost global problems and inevitably conclude with little more than platitudes and empty gestures or market-obsessed plans of action. With a wealth of damning evidence, well-meaning alternative outlooks, such those featured in the Millennium Ecosystem Assessment or the Millennium Development Goals of the United Nations, do the utmost to avoid any discussion of capitalist causes and instead promote palliatives of better 'governance,' poverty-reduction programs, and similar policy recommendations that refuse even to acknowledge the presence of the proverbial elephant in the room. Anti-migrant, blatantly fascist and racist, and unreconstructed misogynistic rhetoric and violence is the diversionary flip side of the same sustained capitalist policies of free-market promotion, leading to the displacement of the costs of profits onto those with the least means, because of current or historical deprivation, to turn the tide in a constructive direction for themselves and the world.

What is refreshing about this book is not only the combination of technical expertise with unabashed politics. It is the culmination of decades of scientific research into energy and farming systems as well as decades of direct involvement in electoral politics. The authors' ideas are therefore hardly coming from armchair warming exercises. They could not be more attuned to the issues at stake and to realistic alternatives. Unlike most scientists in the biophysical sciences, who pretend to be politically neutral or mask their politics by appeals to objectivity, Schwartzman and Schwartzman give a thorough technical treatment on the crucial environmental issues of our times while spelling out their commitments with honesty, laying out and critiquing the social order that conventional science prefers to leave unnamed and unquestioned.

Frankness and directness is what is most needed in times of urgency, and global warming should loom large in all scientists' minds as a problem to confront in ways that result in least harm and in better futures for all.

The importance of Schwartzman and Schwartzman's work could not be more important for another reason, too. Leftist responses have been mixed, often anemic, and overwhelmingly reliant on statements and publications from politically mainstream scientific experts. On the hopeful side, many have recognized the ecocidal nature of capitalism and have organized against capitalist forces, in their various guises of democracy or religious or military dictatorship, even across continents in political actions and by establishing or expanding upon existing egalitarian practices and traditions. But this is contradicted, if not undermined by much of the more politically empowered and economically endowed academic and wider intellectual left, whose self-serving readiness to disown anything socialist (since at least the 1990s) and whose decades-long privileging of critique (including of leftist theory and practice) impedes concerted action by stifling the development of political strategy and practicable solutions. At the same time, institutional leftist organizations and parties increasingly veer to or are compelled to conform to liberal democracy (i.e., the political manifestation of the militarily and economically most globally powerful capitalist forces), if they have not already essentially fused with it. These variegated aspects of leftist responses are intimately linked in that a lack of effective organizing beyond state institutions is reflected in the often right-wing pro-capitalist turn in erstwhile leftist parties.

This is why this book must be read and discussed widely — and soon — even among those who still believe in the status quo or think reforms are sufficient. One may disagree with the eco-socialist politics espoused by the authors, but the evidence presented in this book, both environmental and social, should compel even the most recalcitrant to re-assess how things stand and at the very least to counter with a credible, practicable alternative that goes beyond, as Schwartzman and Schwartzman call it, business as usual. For those who already have made up their minds about the necessity to overcome capitalism, this book ought to become a fundamental resource not only to identify the main sources of the most pressing global problems, but also to devise actionable alternatives. Irrespective of justifiable objections to the depoliticizing effects of catastrophism, as some view these kinds of discussions on the left, there are points beyond which deleterious changes are irrevocable and in many case they have already happened, such as the forced migration of thousands of especially Indigenous Peoples by circumpolar ice melting and the submersion of islands in the Pacific Ocean. Catastrophes have already arrived many decades ago with global warming, bringing death, displacement, and loss of livelihood. It is therefore imperative that actions be taken with much greater urgency, to reduce if not prevent even greater and more diffuse destruction worldwide.

To take action means identifying and developing a thorough comprehension of the problems to face. With great clarity and directness, this is exactly what Schwartzman and Schwartzman accomplish by directing our attention to and by explaining the two main pillars of the many forms of environmental degradation afflicting many of us and other species: harnessing energy and procuring food. But they go much beyond recurring narratives that are useful towards appeasing the powerful and that pathetically miss the mark. Schwartzman and Schwartzman show, in compelling detail, how the twin foundations of our very existence, energy and food, have been turned, as a result of the development and spread of capitalist relations, into harbingers of doom for the many. As they show, following on the footsteps of many before them, solar energy abounds and there is plenty of food to feed everyone, so focusing on the growth in global consumption and population levels is misguided. The problems lie in the kind of energy used, how food is produced, and how both energy and food are distributed and consumed. Many others have been pointing out and critically appraising these major challenges, but few if any have endeavored to formulate a way out by using existing sets of conditions and on the basis of scientific principles. In a dialectical fashion, they see potentials to overcome dependence on fossil fuels by using fossil fuels themselves and to supplant industrial farming with pre-existing and emerging agroecological practices.

By evaluating potentials and existing successes, Schwartzman and Schwartzman have developed a blueprint, which is something many who are interested in changing society continue to shy away from. Perhaps it is part of a justifiably negative reaction to grand narratives or generalizations. But blueprints must not be confused with any such thing, just as envisioning a better future is not the same as foreclosing it. Strategies and plans are as necessary as the striving for the inclusion of egalitarian voices that have been repressed within the Left (and, besides, devising processes of inclusion are simultaneously blueprints for action). This book can therefore be read, among other things, as an outline of a viable solution to the necessary transition away from currently and historically ecologically devastating and socially undermining ways of procuring and consuming energy and food.

As Schwartzman and Schwartzman never cease to emphasize, none of these technical solutions are feasible without tearing down the biggest and most destructive capitalist institutions that stand in the way of making these transitions: the military- and poverty-industrial complexes. To achieve this, every effort needs to be made to spread awareness of where the main problems lie (militarism and systemic deprivations), so as to enable the introduction and implementation of solar energy and agroecological techniques worldwide. Happily, the world is not starting from scratch, and Schwartzman and Schwartzman show us where alternatives have succeeded and how they can help the rest of the world move out of conventional, normalized insanity. Central to prefiguring and definitively establishing socially and ecologically solar futures across the world is a multidimensional class struggle, waged by seeking complementarities and enabling coordinated action among those oppressed. We have a couple of decades before global warming may usher in destruction even greater and more extensive than what is already happening. Yet even if the most dire of predictions are postponed or do not materialize at all, existing catastrophes are and have been more than enough to take action. It must not be the same kind of action as before or that is hammered into us in the mainstream as the only possible. This book exposes the folly of such action and points to the sort that is effective, at this juncture, towards bringing about worldwide social and environmental justice and brighter futures for all and not just for humanity. We call it ecosocialism.

Salvatore Engel-Di Mauro
Editor-in-Chief of *Capitalism Nature Socialism*
Professor of Geography, SUNY New Paltz,
New Paltz, New York.

Acknowledgments

P. Schwartzman wishes to acknowledge the following people: (research support) Howard Ehrman, Sofia Tagkaloglou, Steve Cohn, Konrad Hamilton; (general support) his partner Tricia Zelazny and his two children, Camellia and Juniper, and John Hunigan.

D. Schwartzman wishes to acknowledge the following people: (research support) Quincy Saul, Salvatore Engel-DiMauro, Joel Kovel, Tyler Volk, Michael Rampino, Amelia Amon, Robert Biel, Bill Bowring, Brian Tokar, Richard Greeman, Jane Zara, Peter Caplan, Kurt Stand, Victor Wallis, Walter Teague, Doug Boucher, his colleagues at Howard University, in particular, George Middendorf, John Tharakan, John Trimble, Chuck Verharen; (general support) his partner Joanne Fleming, Emilie Junge and our son Sam Junge, and last but not least Mousey and Porky Pants, daughter and mother, street cats adopted from Baltimore.

About the Authors

David Schwartzman (left) & Peter Schwartzman (right)

Peter D. Schwartzman has taught Environmental Studies at Knox College (in west-central Illinois, USA) since 1998. He holds a PhD in Environmental Sciences (University of Virginia); MSc in Science and Technology Studies (Virginia Tech, USA); BSc in Physics with minor in Philosophy (Harvey Mudd College, USA). He has served as Alderman in Galesburg, IL since 2011, having been elected twice. He has co-founded two locally-focused non-profits (Knox Prairie Community Kitchen and Growing Together, Inc.) and has served as a board member on many others (including, Galesburg Athletic Youth Club, Galesburg Farmers' Market Association, Illinois Stewardship Alliance, and Western Illinois Nature Group). He oversees two websites: solarutopia.org and onehuman.org (personal blog). He has two children and loves to play Scrabble, basketball and tennis.

David W. Schwartzman is Professor Emeritus, Howard University (Washington DC, USA) and is a biogeochemist and environmental scientist. He holds a PhD in Geochemistry from Brown University, USA. He contributes to his older son Peter Schwartzman's website solarUtopia.org. His publications include: *Life, Temperature and the Earth* (2002), and several recent papers in *Capitalism Nature Socialism* (CNS). David serves on the CNS Advisory Board, and is also on the Advisory Board of *Science & Society*, and the Institute for Policy Research & Development. He is an active member of the DC Statehood Green Party/Green Party of the United States as well as several other community organizations, especially since his retirement from Howard University at the end of June 2012.

Contents

Figures and Tables

Introduction

We begin with optimism as we look forward to a future for our own children and grandchildren — indeed all those now living on our planet. We are confident that humanity will find a path to the realization of the immense potential of technologies that already exist to meet human and nature's needs, and we are inspired by the energy and creativity of the youth of the world who will lead and make possible this transformation into the other world that is possible. As John Lennon put it in his song *Imagine*, 'You may say I'm a dreamer, but I am not the only one.'

Nevertheless, we are not naïve. We live in a world of great inequality. Your authors live in the global North and enjoy a much higher standard of living than most people living on our planet. We get a glimpse of the actual state in which most of humanity lives, mostly through our TVs and the web, or more recently, when we see homeless people sleeping in the streets of our major cities and/or immigrants from the global South who have fled their countries because of conflict and deprivation. Yes, even in the global North, a significant fraction of its residents continue to endure a crisis of everyday life, with housing becoming increasingly unaffordable.

One of the authors (David) lives in Washington DC, the capital of the United States, the richest and militarily most powerful nation on Earth. Yet DC has very high income inequality and shockingly high child poverty (Kids Count 2014; Rivers, 2014). The other author (Peter) lives in a city of about 30,000 in west-central Illinois, surrounded by acres and acres of GMOs (Genetically Modified Organisms) — primarily corn and soybeans — which are increasingly being grown to make sugar and gas substitutes. Governmental subsidies continue to go to these crops despite the increasing signs that American diets are poor, and, in particular, children are beginning to suffer from preventable diseases/illnesses (e.g., premature puberty and Type II diabetes). The United States ranks near the bottom of OECD countries with respect to both life expectancy and child poverty rates (Healthy Living, 2013; OECD, 2017).

Close to six million children in the global South still die of preventable causes every year, although underfunded UNICEF has made dramatic strides in reducing child mortality in the last few decades (Garrett, 2015). Adequate funding of UNICEF programs would virtually eliminate preventable child mortality (UNICEF, 2013) — and this would require only a small fraction of the current global military expenditures totaling US$1.8 trillion per year (Perlo-Freeman *et al.*, 2015). In 2014, global conflicts cost US$14.3 trillion or 13% of the World's Gross National Product (GDP), and resulted in 180,000 deaths (BBC Online, 2015). Global inequalities are highlighted by the global US$300 billion per year being spent on luxury consumption, while less than one-third of this spending could fully fund UNICEF programs. Of course, we recognize that the production of luxury goods produces employment for those who are not part of the 0.1–1% of humanity who are their consumers. Nevertheless, given the unmet needs, particularly of the billion people still living in dire poverty, we conclude that the luxury sector could be greatly reduced with a transfer of human and material resources to meet these unmet needs.

In the last few decades, the human impact of climate change has been accelerated. The burning of fossil fuels (and wood, especially in the global South) generates air pollution and carbon emissions to the atmosphere. Here is a snapshot of their effects:

(1) A WHO study found that seven million people died from air pollution in 2012, approximately one-eighth of global mortality (WHO, 2014).
(2) A 2012 study found that 400,000 deaths per year were caused by climate change, with the toll rising if mitigation of global warming is not implemented in the near future (Hertsgaard, 2012).

Just as we were finishing the manuscript of this book, Hurricane Harvey was finally dissipating its energy over the state of Louisiana after unprecedented flooding in Texas, including its biggest city Houston, with a population of over two million. Leading climate scientist Michael Mann wrote:

> 'What can we say about the role of climate change in the unprecedented disaster that is unfolding in Houston with Hurricane Harvey? There are certain climate change-related factors that we can, with great confidence, say worsened the flooding...Sea surface temperatures in the area where Harvey intensified were 0.5–1°C warmer than current-day average temperatures, which translates to 1–1.5°C warmer than "average" temperatures a few decades ago. That means 3–5% more moisture in the atmosphere. That large amount of moisture creates the potential for much greater rainfalls and greater flooding. The combination of coastal flooding and heavy rainfall is responsible for the devastating flooding that Houston is experiencing' (Mann, 2017).

'Extreme weather events are the new normal', says *Nature* in reference to Hurricane Harvey (Editorial, 2017). And other local impacts of Hurricane Harvey are already being witnessed, with the release of hazardous pollutants from oil refineries (Mufson, 2017a) and explosions at a chemical plant (Mufson, 2017b; Bagg *et al.*, 2017).

Our climate is only one 'planetary boundary' being rapidly approached by unsustainable impacts of our global civilization on the environment. According to a recent study, the two other most severe are an accelerating loss of biodiversity, with a sixth mass extinction currently underway (Ceballos *et al.*, 2015; 2017), and the overload of biologically available nitrogen derived from industrially fixed nitrogen released into the environment (Rockström *et al.*, 2009), with both processes linked intimately to climate change. For example, chemical reactions involving fixed nitrogen in the soil drives the release of the potent greenhouse gas nitrous oxide into the atmosphere.

One of the authors (David) wrote a science fiction projection of four scenarios covering the present to mid-century (Schwartzman, 2013). Here they are, minus the references and most of the footnotes:

> 'In the order of increasing likelihood...
>
> Scenario #1: 0.1% likelihood of happening.
> We muddle through. Following the global depression of 2008–2020, and three more cycles of boom and bust in global capitalism that follow, the

pattern of the 20th Century is repeated. The climate remains pretty much the same as today. The current projected impacts of global warming are not realized because of unforeseen negative feedback in the climate system. Even the ecological impacts of rising carbon dioxide levels in the atmosphere and resultant acidification of the ocean are mitigated by unforeseen rapid adaptations of marine organisms, leaving biodiversity and oceanic productivity virtually unaffected… [As of 2018 the global economy is growing again, but given the powerful role of finance capital, there is every expectation that the bubble will burst again, consistent with the boom and bust cycle inherent in the capitalist economy].

Scenario #2: 10% possibility of occurring.
Between 2015 and 2020, large-scale implementation of high-efficiency thin film photovoltaics, low-cost capture of ocean currents, and high-elevation tapping of wind energy begins to rapidly decarbonize global energy supplies, radically undermining the Military Industrial Complex (MIC) because of the growing availability of very low-cost clean energy, which requires virtually no rare strategic metals. Corporate-instigated attempts to block this rapid process of solarization are undermined by decentralized grassroots initiatives around the globe. Massive civil disobedience and resistance within the armed forces and police prevent any effective repression of a now global peace and justice movement fighting for survival in the continuing Global Slump. As a result, public support for the MIC plunges, governments are elected around the world, including in the United States, with anticapitalist agendas, promising a 21st Century ecosocialist transition to Solar Communism. The dreams of Marx and W. Warren Wagar are realized.

Scenario #3: 30% probability of taking place.
(Am I an optimist or a just wishful thinker, given this has one-half the probability of scenario 4?) In the year 2016 [alas didn't happen!], what now is thought impossible happens: The explosive growth of a transnational peace and climate security movement begins as a response to the Great Slump, continuing resource wars, and the escalating impact of global warming...An ecologically oriented Conversion Economy emerges linking most of Africa, China, India, South America and Russia. By 2018 [if only!], the military budgets of the United States, Russia, and China

are reduced by 75%, and these resources are transferred to a global program of climate, food, and health security...The annual genocide of ten million children dying of preventable causes under the global rule of capital is finally terminated by the UN investment of US$80 billion, which is extracted by a Tobin tax on financial speculation.[I.1] Agroecologies inspired by the vision and practice of permaculture blossom in and around cities across the globe. High-efficiency solar power begins to rapidly replace fossil fuels and nuclear power, as well as serve to sequester carbon dioxide out of the atmosphere to reach levels below 350 ppm, the minimum level sufficient for reaching irreversible tipping points leading to catastrophic climate change. *Per capita* energy consumption in the global South increases to the point where the state-of-the-science life expectancy, health, and education are possible for all citizens. The unsustainable consumption in the global North plunges to levels that ensure clean air and clean water, organic food, meaningful employment, and more free, creative time for all on this planet, realizing global equity and the highest quality of life for all.

Scenario #4: 60% probability. Unfortunately, the most likely, but still not inevitable.

The Great Depression of 2008–2020 is terminated by a combination of fascist repression, mass availability of the new inexpensive opiate, with which people drug themselves (the prison industrial complex now enslaves political prisoners, with the drug war being no more) and the emergence of a rigid caste overlay on existing class divisions: One-fourth of the globe's population consumes 95% of the unending flow of commodities, which become obsolescent or dysfunctional in record time. The Military Industrial (Fossil Fuel/Nuclear/State Terror) Complex's dominant role in global capital reproduction and expansion continues unrestrained. By 2025, ecocatastrophe kicks in on a global scale, with the climate model projections made in the early 21st Century now found to be far too conservative. Irreversible explosive releases of

[I.1] Ten million was the UN estimate for the 1970s to the 1980s. A significant reduction in these preventable deaths was achieved by the beginning of the 21st Century, but the misery index of the global South (and even in the global North) grew during the deep and prolonged economic crisis from 2008 to 2017.

> methane occur from melting permafrost (warning signs already apparent 40 years earlier; Connor, 2011) [but see discussion of recent research in Chapter 4] resulting in the complete disappearance of Arctic sea ice and ever-accelerating melting of the ice caps, flooding coastal cities around the globe. Ocean acidification kills off the existing marine ecosystems, starting with coral reefs. This increases the future prospect of a rapid increase in hydrogen sulfide-producing bacteria, which now threaten to poison the whole planet, replicating the end Permian mass extinction. A few thousand billionaires and their families survive in nuclear-powered air-conditioned enclosures, raising organic food in greenhouses. By 2050, most of humanity have perished from starvation and disease.'

We believe that there is still a chance for something like Scenarios 2 or 3 to become reality, avoiding the climate hell of Scenario 4. Drawing on our published research and the extensive work of many others, our book will address these challenges on a global scale, with a synthesis that outlines a clear path to the 'other world that is still possible' ('OWSP').

In Chapter 1 we offer an imperative underlying our vision: To provide the highest quality of life to all humans while preserving biodiversity. We will introduce the critical challenges currently facing humanity with respect to environmental degradation, and in particular what climate science is telling us about the threat of catastrophic climate change (C3) — specifically what is the science behind understanding climate change and its current projections for the rest of this century if a prevention program to avoid C3 is not implemented in time. We address the following: What are the current impacts of the lack of clean energy on climate change, human health and biodiversity on our planet; and, what are the challenges humanity now faces with respect to the availability of nutritious food and the agricultural systems which are assumed to produce it?

In Chapter 2 we will go back to the 18th Century to find the roots of these huge challenges of the 21st Century, with a historical account of the stages of industrial revolution through the 20th Century; how the science of thermodynamics, so critical to understanding energy, was born in the 19th Century and why this science is critical to effectively addressing current energy and food challenges.

Chapter 3 will further examine the why the current modes of energy and food production are unsustainable, and what the huge negative impacts current agricultural production to humans and nature are, with its emphasis on meat supply and use of fossil fuels and nitrate fertilizers.

With the quantity and quality of energy available to our global civilization being foundational to meeting all needs, in Chapter 4 we will answer the question: how much and what kind of energy does humanity actually need? Answering this question will also be the focus why wind/solar energy technologies are the 'coal' of the 21st Century, the analogue of how coal and steam power were the driving force for the industrial revolution using biomass as the energy source to make that transition possible. In this chapter, we will show why the quality of wind/solar power is fundamentally different compared to fossil fuels and nuclear power, and how all fossil fuel consumption can be finally terminated in a robust wind/solar power transition on our planet. We will show that this transition can be completed in just a few decades, in an optimized path to a fully renewable energy world, which simultaneously has the best chance to avoid C3 and ending energy poverty, which now shortens the lives of the majority of humanity.

Chapter 5 will examine in more detail what the dominant food paradigm is, followed by a systematic discussion of the basis for the transformation of agriculture into science-based ecological production (agroecology), and why this transformation is imperative for creating a sustainable food system to ensure the health of people and thriving ecosystems around the world. An effective prevention program to avoid C3 can only be achieved in global regime of profound cooperation, rather than conflict, in a demilitarized world. Thus, we submit that there are three material requirements for the other world that is possible: demilitarization, solarization and creation of ecologically-based farming. The realization of these requirements will require an unprecedented level of grassroots mobilization at every scale from the local to global. Therefore, in Chapter 6, we will further examine why the Military Industrial Complex (the MIC) is a huge obstacle to creating the OWSP. Further, we find both false solutions and false obstacles being articulated that detract from a serious program with the capacity to achieve this transition. The false solutions include the promotion of most biofuels and nuclear power derived from fission. The false obstacles include the mantra that overpopulation is the prime driver of biospheric collapse rather than the gross inequities and unsustainability of the dominant physical and political economies. Further, we will discuss the other ideologies and practices that divide and oppress people, blocking the unity needed to build stronger social movements.

Economic growth is the accepted goal for capitalist economies by a broad spectrum of political leadership. A critique comes from the degrowth discourse, prominently represented in the green environmental movement. In Chapter 7 we will critique this critique by examining the qualitative as well as quantitative aspects of economic growth, confronting the challenges of creating the OWSP. The history of the New Deal will be examined with its relevance towards what we propose as a critical path out of fossil fuel capitalism, the Global Green New Deal.

Cuba is at the cutting edge of implementing agroecologies, as well as making remarkable strides in health and education. Hence, we will address in Chapter 8 why the Cuban experience should be taken very seriously in considering what a global *ecosocialist* transition will entail (in a nutshell ecosocialism is a socialism compatible with the ecology of the planet). Likewise, we address how Venezuela can potentially lead this transition in Latin America. And we reflect on the remarkable energy transition to renewables taken by Germany in the last few decades, the Energiewende, and what lessons can be learned to bring power to the people, democratizing and socializing the management and control of energy production. China, a huge player in the future direction of energy, is discussed as well.

Finally, in Chapter 9, we conclude by demonstrating why only the multi-dimensional, transnational class struggle in the 21st Century can deliver the OWSP, snatched out of the chaos that now confronts us in the converging environmental, social, economic crises. We will confront the necessity and challenge of moving beyond capitalism, and address how realizing the potential of abundant energy can make possible an ecosocialist transition. We will draw from the valuable resources of 'The Next System Project' and insights from Jeremy Rifkin's recent books, The *Third Industrial Revolution* and *The Zero Marginal Cost Society*. Value is a concept in Karl Marx's *Grundrisse* and *Capital*, central to his critique of capitalism. We will discuss its relevance in the 21st Century in the context of an ecosocialist transition. We will explore the powerful concept of prefiguring the future, which means creating concrete models of living differently while embedded in an unsustainable global system. Prefigurations of the OWSP already exist and the future must be informed by them as well as cutting edge scientific theories in thermodynamics, biogeochemistry, information science and agroecology. We aim to provoke, which should be obvious by now.

Chapter 1
Our Current Situation

1.1 What key challenges face humanity?

Many recognize the 21st Century is probably a make-it or break-it period for our species. While it might seem implausible that we could wipe out a substantial portion of the global population and radically change biodiversity as we know it on such a short time scale, it is a possibility that we must take seriously. In order to tackle a problem/dilemma of such grand proportions, we must consider ultimate (distal) causes of this potentiality (and, further, distinguish them from proximate ones). With this intent focus in mind, we ask, would we not be better positioned to come to a collective awareness of our future and make the necessary changes so that our civilization averts a multitude of catastrophes?

Before we begin, let us acknowledge that the above is couched in dire terms. Yet, before readers dismiss this work based on this position, as dismal diatribes are a dime a dozen these days, let us point out that we, the authors, do not believe that fear serves as a good motivator for positive change. However, it is not sensible to ignore the gravity of our situation either. We, thus, carry forward with this project with the mindset that humans have the capacity to create a much better future than the reality we now witness.

What are the ultimate causes to the threat of our species impending demise? The threat of climate change is an opportunity to not only avoid mass destruction but to create a flourishing integrated and harmonious society. The obstacles? Capitalism, materialism, imperialism, inequality, overpopulation (which we soon debunk), toxic chemicals, and militarism all come to mind. Racism, sexism/misogyny, xenophobia, homophobia, and cultural chauvinism also deserve mention as complementary factors in such a list. Determining which of these are proximate or ultimate causes of our potential 'doom' would be a book unto itself, yet some discussion of them is mandatory if we are going to arrive at potential remedies and solutions. As the saying goes, you cannot fix a problem unless you first understand what is causing it.

At least that is how a logical (rational) person might address these challenges. However, paradoxically, for our purposes, we may not need to know for sure what will be the ultimate genocidal agent. Likely, a multiplicity of agents working synergistically in nefarious ways will be our undoing. In fact, it may be that focusing on 'doom and gloom' (as we find occurring in most media outlets these days, both on the right and left of the political center) merely distracts us from more meaningful, purposeful and productive engagement. What if we began to focus on goals first rather than on 'all the problems' or 'all the disharmonies'? Might we, by so doing, refocus/redirect feelings of despair, hopelessness and frustration towards positive-orientated thoughts about capacity, opportunity and progress? What indeed are the collective ideals to which we aspire? And, can these be reached while at the same time avoiding continued global degradation? Notice, if the answer to this last question is 'no', then we must return to our goals once again. One might like German chocolate cake the most but when one considers the health impacts of eating it exclusively, one reconsiders if 'eating blissfully' is an appropriate goal.

Prioritizing goal setting does not make our work/project any easier. With plenty of options for guidance, there exists one theoretical construct which appears rather compelling as we enter into this radical transition of civilization as we know it — the precautionary principle (PP). Its tenets dictate that we need to proceed with caution when we engage in behaviors/activities that we have reason to think might be harmful to humans, and life more generally. It demands that we consider all the options available to us and that we continually assess and reassess decisions made (and actions taken) to determine if better ways exist to achieve our collective ends.

The PP has been successfully used to craft many policies at the global, national, and local levels. The Science & Environmental Health Network (founded in 1994) has dedicated a lot of work on both the theory and operationalization of PP (an extensive FAQ on PP can be found on the organization's website: SHEN, 2017). In the context of PP, the ends serve as the goals, and establishing/setting goals becomes a fundamental and commencing act. Humanity could benefit greatly from a deep reevaluation of its goals, something akin to the Universal Declaration of Human Rights (passed 70 years ago in Paris) or the Rights of Mother Earth (passed in 2010 by the Bolivian legislative assembly). Along these lines, and for the purposes of furthering discussion, we offer the following as humanity's goal of the 21st Century:

> To provide the highest quality of life to all humans while preserving biodiversity.

Recognizing, of course, that other equally desirable goals could be formulated, our goal is not to pick the 'best' one but rather to show how one such goal could act as the organizing principle guiding the grand transition that we all need to make. Other steps exist in the operationalization of the PP (including, 'exploring *alternatives* to possibly harmful actions', 'placing the *burden of proof* on proponents of an activity rather than on victims or potential victims of the activity', and, 'bringing *democracy and transparency* to decisions affecting health and the environment' (PP FAQ, n.d.) and these considerations need consideration as well, but here, identifying the goal is sufficient for our purposes. The remaining chapters of this book focus on identifying viable solutions to our pressing problems and indicating how they are being activated as a prefiguration, a provocation for others to evaluate them in theory and in practice.

Simple but effective, our goal makes it clear that all humans are relevant to any actions taken. On this score alone, most capitalistic activities would be deemed misguided, as maximizing profits necessitates exploitation and oppression. The precept to 'preserve biodiversity' demands that we don't '[pave] paradise and put up a parking lot' (as expressed by Joni Mitchell in the 1970s). It also implies the need to eschew hubris as it relates to an understanding of what is 'best' for biodiversity, e.g., many technological fixes have made things worse not better. In other words, the challenge facing humanity this century is to find ways to live in harmony with the biosphere or suffer from its continued breakdown. The fact that most of us (in the USA at least) act as if nature is nothing more than an inconvenience (such as roaches and mice in a house, mosquitos or flies on a patio, 'weeds' in our gardens or on a lawn, or squirrels and opossums on a roadway) suggests how far we must still go. More optimistically, thousands of national and international non-governmental organizations (NGOs) have been formed to work on this challenge (see Paul Hawken's *Blessed Unrest,* or his short video, for an expansive list; Hawken, 2007). And while this ecologically-centered movement manifests and should, if fully actualized, ward off the most damaging and destructive activities, it takes as its first premise that the goal is to focus on our core aspiration — to enhance the lives of humans to their fullest.

How does one provide the highest quality of life? What does this mean? Is it not strange that this question does not have clear and unequivocal answers, given that humans have been on the planet for over 200,000 years and so many now claim that we live in a civilized society? In fact, does this not essentially show how uncivilized modern civilization really is? Another way of asking the question is to focus on the requisite needs for the self-actualization of all. In other words, what do humans need to thrive (rather than merely survive)? One widely-used measure of, or proxy for, 'thriving' is longevity. Amartya Sen's seminal article, 'The Economics of Life and Death' (1992), succinctly demonstrates the power of this measure. Sen's metric of long life enables a rather simple, and fairly effective way to measure human success. But, what does Sen conclude is necessary for a long life? Might it be affluence and economic growth?

Looking at longevity worldwide, much is revealed. While it is true that most nations that have high life expectancies are among the wealthier ones, there are several important exceptions that seem to suggest that excessive wealth is not necessary for obtaining a long, healthy, and happy life. Our analysis of 129 nations with a population of four million or more (with an average life expectancy of 71.0 years and GDP/capita of US$12,185) reveals that only one country with a life expectancy (LE) of 79.5+ years (i.e., at least one standard deviation above the mean) has a below average GDP per capita: Costa Rica (World Bank, 2015b, 2015c; WHO, 2016); see Fig. 1.1 for visual of this info (notice that Cuba and the U.S. are both just barely below this LE threshold, though Cuba's GDP per capita is 14% of the U.S.'s). A few of the more populated countries (namely, China, Mexico and Vietnam) have life expectancies of over 76 years despite all having below average GDP per capita values. Note China and Vietnam had

Figure 1.1. GDP/Capita (US $) versus life expectancy (years) (Nations with populations of 4 plus million)

Sources: GDP/Captia → World Bank (2015c); Life Expectancy → WHO (2016).

socialist economic systems (although private capital is now ascendent), while Costa Rica has eliminated its military. Figure 1.1 also reveals that beyond some level of GDP per capita (~US$10,000), average life expectancy is almost guaranteed to be above 70 years. However, the United States; one of the wealthiest countries in the world, is far from exemplary in terms of longevity. The U.S. also does particularly poorly in the area of infant mortality (Strauss, 2016). All of this is more remarkable when one considers that the U.S. spends far more per capita on healthcare than all other nations, with this disparity being quite substantial when compared to nations such as Japan (the U.S. spends more than twice as much as Japan) which have the longest life expectancies in the World (Brink, 2017). As a result of all this information, we conclude that excessive affluence is not a prerequisite for reaching our ultimate goal. Notice that since the significance of GDP is exaggerated in 'highly developed' countries as compared to 'less developed' ones because it does not account for (non-paid) volunteer and housework, as well as unsustainable production, this important conclusion is only strengthened. If affluence (as a result of continuous growth of economic systems) is not sufficient or necessary for a healthy global community, what is?

According to Sen (1992), the keys to living long lives include access to healthcare (particularly, prenatal and infant) and education, gender equality, and food security. In Sen's analysis, it appears necessary to focus on social and political programs, as it is precisely these that dictate who has/gets health care and food; Sen points out that no democratically-led country has ever endured a famine. Within the US context, focusing on health and education would appear to require an exorbitant expenditure of resources, restricting such reinvestment to the most well-to-do countries. However, such an assumption, which we suspect is deeply imbedded in most contemporary policy makers and bureaucrats in the capitalist world, is ridiculously unfounded. Actually, as Sen (1992) cogently argues, both health and education are labor-intensive enterprises, particularly in the areas that are most important — prenatal, infant, and youth — rather than capital-intensive ones. Also, much of the horrific death of the young from contagious diseases and contaminated water can be eliminated with relatively small expenditures in vaccines, antibiotics, and ceramic water filters. China and Kerala (southwestern state in India, population 35 million in 2012) demonstrate how impactful small, yet strategically invested, contributions can be (Sen, 1992); Cuba, though not explicitly highlighted in Sen (1992), has shown great success in this regard as well, weathering their "Special Period" without disastrous consequences (more on this later). Thus, based on these noteworthy examples, the vast majority of countries in the world are now capable of bringing much improved educational and health-related services to their citizens despite their relative economic disadvantage. Reflecting on these revelations, we conclude that humans are in a good place regarding their potential to meet critical human needs of healthcare and education, two core prerequisites for a thriving civilization.

Returning to our stated 'Humanity Goal', and following Sen's prescription, we need to do everything in our power to maximize access to healthcare (particularly, prenatal and infant), high quality education, gender equality, and food security. Providing state-of-the-science healthcare is not difficult if one makes it a priority. Cuba, a poor nation in terms of GDP per capita, serves a great example of what a concentrated effort to improve healthcare can do. Educating a high percentage of its citizens to enter the healthcare profession (including doctors) allows Cuban citizens to seek out most medical consultations and procedures without the huge expense and duress found in wealthier nations; Cuba's current doctor–citizen ratio is the highest in the world (7.5 physicians per 1,000 citizens; U.S. compares with a paltry 2.6) (WHO, 2017b). As a result, Cuba's infant mortality rate is lower (four) than the U.S.'s (six deaths per 1,000 live births) (World Bank, 2017); this statistic shows as much as any other how challenged the U.S. is with regard to its expensive healthcare system. Kerala, India, has also shown the power of prioritizing pre-natal and maternal health, occupational and educational gender equality to improve health statistics of its population (Sauvaget *et al.* 2012). Remarkably, we find this to be true in both 'rich' as well as 'poor' regions. The projected global distribution of chronic disease deaths is spread out over countries of all economic classes: 35% in low income countries, 45% in middle income countries, and 20% in high income countries (World Bank, 2005). According to WHO, 'at least 80% of premature heart disease, stroke and diabetes' and 'over 40% of cancers' can be prevented by eliminating known risk factors for these diseases (WHO, 2017a). Importantly, these 'interventions are cost-effective for all regions of the world.' (WHO, 2017a). These risk factors are primarily: (a) unhealthy diet; (b) physical inactivity; and (c) tobacco use. Because of the dominance of these risk factors, each year: (a) 4.9 million people die from tobacco use; (b) 2.6 million die due to being overweight or obese; (c) 4.4 million die of raised cholesterol levels; and, (d) 7.1 million die of elevated blood pressure (WHO, 2017b). And interestingly, while deaths from 'infectious diseases, maternal and perinatal conditions, and nutritional deficiencies combined are projected to decline 3%' from 2015–2025; 'death due to chronic diseases is projected to increase by 17%,' which means some 41 million people dying from chronic disease (WHO, 2017b). Clearly, human health suffers from misguided practices in the areas food/diet, exercise, and healthcare. Yet, we would be remiss to fail to acknowledge the tremendous harm done by components of international 'aid', delivered post-World War II, which has burdened many 'less developed nations' (nearly all of Africa) in excessive debt to banks and governments of the global North and, therefore, unprepared to address unmet human needs.

And while Sen's framework has a lot to offer, it fails to adequately address the corruptive and exploitive components of capitalism or the critical impact of modern energy inequality. We contend that capitalism is unsustainable as an economic and political system. It commodifies nature and humans alike and creates inherently unfair and divisive social relationships among peoples. In Chapters 7 and 9, we will examine this further, but here, suffice it to say, capitalism will profit-run a healthcare system unless compelled by sufficiently powerful social movements to do otherwise.

Sen's informative analysis notwithstanding, it should be obvious that humans also need basic resources in order to thrive. No human on planet Earth can currently live a healthy existence without the use of clean water, clean air, adequate energy resources, and nutritious food. Thus, part of meeting our goal would entail providing these essential needs. How do we stand currently on providing these to the planet's 7.5 billion people and are there related challenges faced in efforts to provide them?

1.2.1. *Clean water*

Regarding clean water, we are not doing very well. According to the CDC and WHO, 780 million (more than 10% of the global population) do not have access to drinking water from improved sources, meaning that they now access water from surface sources or other unprotected sources that are much more likely to be heavily polluted and/or disease-ridden. This is not surprising when even in the US, 55% of waterways are considered 'poor quality' by the EPA, meaning that 'they are not clean enough for healthy recreation, public drinking water and subsistence fishing' (Van Noppen, 2013). Disease spreads widely due to insufficient sanitation in many areas. Worldwide, 2.5 billion (~35% of the World's population) 'lack access to improved sanitation' (WASH, 2017), which ensures separation of human excrement and human contact. Sadly, 2,200 children each day die of diarrheal diseases and 88% of such deaths result from 'inadequate availability of water for hygiene, and lack of lack to sanitation' (WASH, 2017).

As a result of inadequate supplies of fresh water resources owing to drought, unsustainable practices and commodifiation, many have projected that the 21st Century will be replete with 'Water Wars'. Michael Klare's 2001 book, *Resource Wars*, carefully outlined the shrinking availability of basic water resources in many parts of the world and predicted that the majority of African countries would soon have resources below basic levels (on a *per capita* basis). The root causes of the current wars in Yemen and Syria include water scarcity driven by drought conditions (Parker, 2016).

1.2.2. *Clean air*

Clean air is becoming more and more difficult to obtain particularly in the metropolises of the developing world. Currently, many of the largest cities in the world have air quality considered harmful 10–30% of the time, see Table 1.1 (Kornei, 2017). Notice that several of these are in the 'most developed' countries, so the problems of air quality are still alive and well worldwide, despite many improvements in air quality in the U.S. over the past 40 years as a result of regulations made by the Environmental Protection Administration (EPA). Urbanization of the World population,

Table 1.1. Unhealthy air in the World's major cities (percentage of time unhealthful in ozone or ammonia levels)

City	% of time ozone levels harmful	% of time ammonia levels harmful
Bangkok	14	16
Beijing	21	20
Buenos Aires	1	29
Delhi	26	74
Dhaka	17	52
Houston	14	8
Istanbul	13	1
Karachi	32	11
Lagos	21	34
Los Angeles	20	0
Mexico City	11	1
Paris	16	7
Tokyo	12	3

Source: Kornei (2017).

which has grown from 33% in 1960 to 54% in 2015 (World Bank, 2016a), only exacerbates the health impacts of urban air pollution on human subjects. All of these 'harmful' days (as well as additional ones that are highly polluted but designated less hazardous than 'harmful') add up to do a great deal of damage. Consider that some 600,000 children are being killed each year by air pollution (Ellis, 2016) and over 3 million people die prematurely as well (Choi, 2015). This crisis stems not only from outdoor air, but also as a result of economically-challenged people burning wood and coal in their homes for the purposes of generating heat and cooking food. Zhang and Smith (2003) estimate that the loss in disability-adjusted life years due to indoor smoke exceeds that of other more highly profiled hazards, such as obesity, lead, urban outdoor air, and climate change. The World Health Organization (WHO) estimates 7 million premature deaths in 2012 as a result of air pollution, including a significant contribution from indoor smoke (WHO, 2014). Nitrogen oxides in air pollution (NOX) are causing immense damage to people and wildlife (Carter 2017). The impact of ozone damage on ecosystems is another major issue that is also well-documented. Dr. Devra Davis (2002) provides an excellent introduction to the history of air pollution in her book, *When Smoke Ran Like Water*.

1.2.3. *Energy*

What about energy? We contend that energy poverty is widely overlooked as a critical variable in the self-determination and self-actualization of a community and its people, whereas in contrast, many prominent organizations largely owe their existence to their focus on food security (e.g., UNICEF, World Food Programme, The Hunger Project, Food and Agriculture Organization (FAO), Feed the Children, Heifer International, etc.). While we will go more into this in later chapters, let it be said now that energy is foundational to addressing the needs of society. For example, energy can be used to access (abundant) ocean water (for desalination), to transport water (and other renewable resources) to areas where it is needed, to grow food and medicine (even underground, as is the case for London's World War II air raid shelters (Carrington, 2016)), and to communicate limitless information and knowledge (through cyberspace).

Without sufficient energy, communities are forced to use only those resources that are immediately available (which becomes more and more difficult to access as finite resources become exhausted or compromised). In contrast to food resources, sufficient energy provides a community with many more opportunities for improving their quality of life, e.g., the printing of books, the lighting and heating of schools/hospitals, the transportation of essential goods, etc. And perhaps most importantly, stated above but bearing repeating, food can be produced with energy (whereas the opposite is not generally true, and in cases where humans make it happen, e.g. ethanol production, as we will also discuss later, it is usually not constructive to moving societies forward). And while sufficient sources of decentralized and non-toxic renewable energy will serve us well in all aspects of our lives into perpetuity, since humans haven't yet built up this infrastructure, it is imperative that we act swiftly to ensure that we use our 'non-renewable' resources very wisely as they will be required to jump-start the initial transition.

Regarding energy, as opposed to food, we actually do not yet have enough energy in production. According to International Energy Association (IEA), global energy supply was 156 PWh (petawatt-hour) in 2012. This converts to 2.5 kW of power use per person. This, according to Smil (2008), falls 30% short of the 3.5 kW per person that is required for all of humanity to get out of energy poverty (and since the affluent members of humanity use well above 6 kW per person, this 30% is an underestimate of the extent to which we currently fall short for most of the global population). So we need more energy than we now produce, not less (as some argue), a conclusion discussed in more depth in chapter 4. And, more importantly, we need different energy sources than we now use. Obvious as it probably seems to those reading this, the fact that we need more and different energy is clearly lost on many countries, investors, and leading politicians. Thus, we feel obliged to provide our documented rationale for this 'obvious' point.

We need to shift from the fossil-fuel dominant energy system of the present to a 100% renewable energy system swiftly and fully. There are four key reasons why we must do this. First, inherently, fossil fuels are very polluting in nearly every aspect. Second, fossil fuels are limited in availability, and we will reach these limits on time scales of human generations (not millennia), particularly if we do little to curb consumption of them. Third, catastrophic climate change (C3) will kick in first before these limits are reached (assuming clean energy doesn't replace them quickly) and massive suffering will ensue. And fourth, fossil fuels are distributed unevenly about the planet, with few nations having immediate access to sufficient quantities. This fact makes them potentially a continual source/force of friction and hostility.

1.2.4. *Fossil fuels and climate change*

Pollution has been a byproduct of using fossil fuels from the onset of their use. Combusting them (aka, burning them in the presence of oxygen) results in lots of products, including carbon dioxide and water. Carbon dioxide (CO_2), a relatively innocuous gas that animals breathe out during respiration, becomes a pollutant when it begins to interfere with the global energy balance via the greenhouse effect. Since the dawn of the industrial age, the burning of fossil fuels has led to the increase of atmospheric CO_2 levels from some 280 parts-per-million to 410 parts-per-million (one can check the current level at CO2Earth, 2017). This 45% increase over about 250 years is the major driver in the enhanced greenhouse effect which has warmed the planet about 1°C over this time (IPCC, 2007). It is worth noting that water vapor, a strong greenhouse gas that is responsible for much of the natural greenhouse warming found on Earth, is also created during fossil fuel combustion as well as evaporation, although this contribution is negligible because water cycles quickly, driven mainly by evaporation from the ocean and precipitation. Recent scientific studies now suggest that water vapor increases from a warming climate, and associated cloud changes, are further increasing global temperatures (Abraham, 2015a). Business-as-usual (B.A.U.) scenarios of continued burning of fossil fuel reserves anticipate that atmospheric CO_2 will rise to levels on the order of 1500 ppm over the next few centuries (NASA, 2017). Changes of this order will have catastrophic effects on the planet, including major modifications of ecosystems, storm intensities, and coastlines (where hundreds of millions of humans still live).

Methane, a greenhouse gas some 28 times more effective, molecule for molecule, at absorbing the Earth's infrared light than CO_2, constitutes the bulk (~75%) of the composition of natural gas. Thus, when natural gas (NG) is incompletely burned or leaks from distribution pipes or holding containers, methane goes directly into the atmosphere. In 2015–2016, a huge NG leak near Los Angeles, California was discovered. An estimated 220 million pounds of NG leaked before it could be permanently sealed approximately four months later (McGrath, 2016)! With increased efforts to extract NG, atmospheric concentrations of methane become more concerning. Recent research also establishes that as the permafrost regions of the world undergo melting, more methane is released into the atmosphere (Abraham, 2015b), perhaps promoting catastrophic consequences (Anthony, 2009; UNEP, 2012). The scope of this positive feedback (climate change serving as a generator for more climate change) remains quite uncertain but could be larger in effect in the 21st Century than the climate effects of CO_2 itself. Historically, methane has had about one-fourth the climate impact as CO_2, despite being in concentrations of roughly 0.5% of CO_2 in the current atmosphere (Blasing, 2016). Hence, methane emissions must be seriously considered in the ongoing IPCC deliberations (Crill and Thornton, 2017).

Though most policy discussions focus on curbing carbon dioxide and methane emissions, other GHGs must also be considered as well. Nearly all have greater impact, molecule for molecule, than CO_2, due primarily to the specific portion of the infrared light spectrum that they absorb. For example, chlorofluorocarbons (CFCs), which got a lot of deserved attention as a result of their involvement in stratospheric ozone depletion and a resultant increase in ultraviolet surface intensity, impact the environment very heavily despite their low concentrations, measured in parts per trillion. Collectively, changes in halocarbons (the CFCs and their replacements, HCFCs, as well as CCl_4) have had about 17% the radiative forcing (a measure of the increased light energy returning to Earth) that CO_2 has had since 1750 (Blasing, 2016); CFCs didn't exist until the 1930s (when Du Pont began producing Freon in large quantities) and HCFCs not until the 1970s, when they were phased in to replace the more problematic CFCs. Nitrous oxide (N_2O), also known as laughing gas, and ground-level (tropospheric) ozone production (a photochemical product of various industrial and transportation pollutants in the presence of heat and light) have also both had appreciable impact on climate forcing — N_2O's impact is about 40% of methane's and tropospheric ozone is 80% of methane's (though it has a much shorter atmospheric lifetime — measured in days versus 12 years for methane). Moreover, nitrous oxide emissions are likely to increase in a warmer climate (Griffis *et al.*, 2017). Nitrous oxide emission hotspots include drained peatlands converted to agricultural use in warm climates (Parn *et al.*, 2018). As such, the restoration of peatlands is also a big potential climate mitigation approach (Leifeld and Menichetti, 2018).

It is worth mentioning that some pollutants, particularly particulate aerosols have a substantial "cooling" impact on Earth. They do so as they are in a solid or liquid state and, therefore, actually reflect sunlight back to space (during the daytime) and enhance cloud reflective properties as well. However, these combined cooling impacts only reduce the radiative forcing components of the GHGs by ~30%, though there is still considerable uncertainty in these estimates (Myhre *et al.*, 2013); some in the 1990s argued that this 'cooling' effect would 'magically' offset GHG forcings — such predictions have not been confirmed, as warming continues.

Based on a variety of possible scenarios, IPCC models predict, global temperature changes of +0.3–4.8°C for the 21st Century (comparing temperatures during the period 2081–2100 relative to 1986–2005), with larger changes anticipated in polar and high-latitude regions (IPCC, 2013). According to NRC (2011), each degree Celsius of global temperature increase can be expected to produce:

- 5–10% changes in precipitation across many regions
- 3–10% increases in the amount of rain falling during the heaviest precipitation events
- 15% decreases in the annually averaged extent of sea ice across the Arctic Ocean, with 25% decreases in the yearly minimum extent in September
- 5–15% reductions in the yields of crops as currently grown
- 200–400% increases in the area burned by wildfire in parts of the Western US

In addition, a 3°C increase would likely see a loss of 250,000 km^2 of wetlands and drylands (NRC, 2011). Importantly, most models used today do not quantify the methane feedbacks, discussed earlier, as they are considered too uncertain at this point to estimate accurately. Additionally, these predictions are tentative and incomplete because many impacts remain difficult to quantify — such as other feedbacks (e.g., those caused by reduced Arctic/Antarctic ice) and methane fluxes from deep-sea sediments (aka, methane hydrates or clathrates). More recently, Henley and King (2017), modeling several scenarios (based on variations of the Interdecadal Pacific Oscillation (IPO)), predict that the global mean surface temperature (GMST) will reach 1.5°C (the maximum target articulated in the recent Paris Agreement of 2016) somewhere between 2024 and 2028. Thus, continued fossil fuel use will result in GHG additions to the atmosphere which only put humans in a more precarious and uncertain position. We will revisit the requirements to keep potential warming below 1.5°C in Chapter 4.

Despite this evidence for impending C3, many scholars act as if a B.A.U. scenario is still realistic (see Shafiee and Topal, 2009; US EIA, 2017). They either ignore likely shortages in oil from the future depletion of reserves or predict that current (and even increased) levels of material consumption will continue until nearly all fossil fuels are exhausted. In either case, several very important considerations are overlooked. First, if peak fossil fuels occurs soon (with peak referring to the global production level), peak 'oil' will most likely come first (as predicted by most energy experts based on the known reserves and assumed rates of consumption, with coal having greatest reserves, e.g., see Shafiee and Topal, 2009; Hansen *et al.*, 2013, Figure 2.). The use of conventional oil for energy produces less GHG emissions than other fossil fuels. Therefore, assuming no alternative source of energy, peak oil followed by its reduction in production would lead to increased usage of coal or natural gas as substitutes, thereby accelerating GHG emissions. Second, again assuming a B.A.U. dependence on fossil fuels and neglecting the climate change impacts, if peak oil occurs in the near future, it will likely result in great social and economic instability, especially given how unevenly fossil fuels are distributed across the planet, i.e., few of the World's 200 plus nations have sufficient quantities of fossil fuels to be self-sufficient, with the United States being largest importer of petroleum in the world. Many 'rich' countries currently rely on energy imports to function, as illustrated in Table 1.2. Of the 49 countries (with populations of at least four million) that

Table 1.2. Net imported energy (as % of total energy used)

Net importers	**Net imported energy (% of use)**	**Net exporters**	**Net imported energy (% of use)**
United States	7	Russia	–84*
France	44	Indonesia	–103*
United Kingdom	35	Canada	–73
Germany	61	Mexico	–5
Italy	76	Australia	–190
Japan	93	Nigeria	–93*
Israel	65	Saudi Arabia	–192*
Brazil	12	UAE	–184*
India	34*	Oman	–206
China	15*	Iraq	–229
		Iran	–33*

Source: Compiled from data found at World Bank (2015).
* indicates 2014 data, otherwise 2015.

have a GDP per capita greater than average (>US$16K), only twelve (or <25%) are net exporters of energy; and of the ten most populated countries (that collectively constitute 58% of the human population), only three are net exporters (Indonesia, Nigeria, and Russia) (World Bank, 2015). In addition, taking into account the very strong evidence for a link of continued fossil fuel use to the threat of C3, we submit that it is truly absurd to model B.A.U. scenarios as realistic.

Hence, if one considers C3 as a serious threat driven by continued use of fossil fuels, one recognizes that the only realistic future that avoids calamity is a rapid transition to 100% renewable energy. As far as a prediction regarding how close we are to peak oil, ExxonMobil and Chevron (the two biggest U.S. oil companies) say it is not in the immediate future while large European producers claim that it 'could emerge as soon as 2025 or 2030' (Cook and Cherney, 2017). One of the world's largest banks, HSBC, predicts a major economic crisis driven by 'peak oil' occurring in 2018 (Ahmed, 2017a; see Fustier *et al.*, 2017, for the HSBC report). We are dubious about such predictions. In any case, those institutions most responsible for global oil addiction are not yet ready to agree to an agenda consistent with preventing C3. This realization points to the incredible importance of engaging with this impending catastrophe immediately (see Jacobson and Delucchi, 2009, and McKibben, 2017). *We need to 'peak' coal and natural gas as soon as possible, followed by their rapid phase out as a means of C3 prevention* while using the minimum of remaining conventional oil, the fossil fuel with the lowest carbon footprint, as a bridge to a sustainable 100% solar energy economy (see discussion in chapter 4). To continue to contemplate scenarios of material consumption on a fossil fuel-based economy and ignore real solutions that will avert the worst aspects of C3 is not only gravely foolish, but a prescription for inevitable chaos and suffering worldwide with civilizational collapse a real possibility in this century. Implementation of the concept of intergenerational equity (Treves *et al.*, 2018) must also be front and center in C3 prevention.

1.2.5. *Other air pollutants from burning fossil fuels*

Other pollutants from combusting fossil fuels have more immediate impacts, as many are very toxic to life when they, or their byproducts, are ingested or inhaled. For example, coal combustion leads to the release of mercury, cadmium, arsenic, sulfur dioxide, and nitrogen oxides, just to name some of the more pernicious (and better understood) products. A typical uncontrolled coal plant emits approximately 170 pounds of mercury each year (UCS, 2017). Mercury is extremely toxic with detrimental health effects observed at microgram levels of inhalation/absorption. In 2013, an estimated 366,000 premature deaths in China were attributable to the burning of coal (Wong, 2016); a study by the Clean Air Task Force estimated that emissions from coal-fired power plants resulted in 13,000 premature deaths in the USA (Schneider and Banks, 2010). Gasoline combustion, particularly in high compression engines, results in the incomplete burning of carbon which produces dangerous compounds such as nitrogen oxides and carbon monoxide as well as particulate matter. And, in the presence of sunlight and heat, these compounds react and produce the highly toxic gas, ozone, as a photochemical byproduct. The American Lung Association estimates that an astounding number of people in the

US, nearly 45%, 'live in areas with unhealthful levels of ozone' (ALA, 2014).

1.3. Climate refugees

Additionally, climate changes of the past few decades have created a new type of refugee. *Climate Refugees* (2017) documents the fact that people have been displaced by 'climatologically-induced' environmental disasters which in its earliest of stages, has already impacted at least 48 of the World's nations. The UNHCR, the UN's Refugee Agency, estimates that some 22 million people have been 'forcibly displaced by weather-related sudden onset hazards — such as floods, storms, wildfires, extreme temperature — each year since 2008', adding that 'no region is immune from climate change' (UNHCR, 2016). Northern Africa and areas of the Middle East are likely to see a dramatic increase in oppressively hot days (Lelieveld *et al.*, 2016), almost guaranteeing a migration of grand proportions. The recent unbelievable 165°F (74°C) temperature observed in Bandar Mahshahr (an Iranian coastal city on the Persian Gulf) may become more commonplace. Political turmoil like that found in Yemen and Syria recently, as previously mentioned, may be a consequence of these climatic changes. Ahmed (2017b) has a valuable discussion of the links between climate change and political violence. We will revisit the projected impact of unconstrained global warming on heat stress in Chapter 4.

1.4. Food

Beyond energy but closely related to human health is the food which people consume. Many commonly believe that modern agriculture systems have done wonders to provide healthful food. Yet, statistics tell a different story. Over a billion (nearly 1 in 7) humans are hungry and two billion (nearly 1 in 3.5) suffer from micronutrient deficiencies (Nierenberg, 2017). These stark statistics exist despite the incredible efforts over the past 70 years in the area of industrial agriculture and, more recently, genetic engineering to improve nutritional content in grain and improve production yields per acre. While percentage-wise, fewer humans suffer from hunger, great challenges remain to ensure that all are provided this most basic of needs. Additionally, overconsumption by increasing numbers of humans also provides a challenge as medical costs skyrocket in more affluent areas associated with expansions in the incidence of chronic illness. Chapter 3 will clarify why these food-related problems persist and examine the scale of their dysfunction and ultimate cost to society.

Given that so much of the planet's landscape has been dedicated to food production (36% of potentially arable land (Bruinsma, 2003)), it raises the question, 'Is enough being left for the other millions of species on Earth'? The answer to this oft-neglected question is 'definitively no'. Modern agriculture is a key contributor to the increasing prospect of a '6th Great Extinction' (Ceballos *et al.*, 2015; Kolbert, 2015; Ceballos *et al.*, 2017). The Center for Biological Diversity (CBD, 2017) notes that we are currently losing somewhere between 1,000–10,000 times the background extinction rate of life forms. At current rates, we may see the loss of 30–50% of species by 2050. Now while this seems very sad to those who view biodiversity as an intrinsic value, it is very meaningful for its potentialnegative impact on human health and survival as well. A special 2015 report put together by a partnership of the UN Environmental Programme, the Convention on Biological Diversity, and the World Health Organization, entitled, 'Connecting Global Priorities: Biodiversity and Human Health', speaks directly to these effects. Humans benefit greatly through the 'symbiotic microbial communities present in our gut, skin, respiratory and urino-genital tracts' (Connecting, 2015). The biodiversity of the biosphere 'underpins ecosystem functioning and the provision of goods and services that are essential to human health and well-being' (Connecting, 2015), a biodiversity which includes pollinators, decomposers and natural enemies of pests. The services provided by the biosphere are vast and include: 'food, clean air and both the quantity and quality of fresh water, medicines, spiritual and cultural values, climate regulation, pest and disease regulation, and disaster risk reduction' (Connecting 2015). Pollinators such as birds, bees and bats, affect 35% of the World's crop production — 'increasing outputs of 87' of them (FAO, 2017b). Thus, recent declines in species richness of pollinators, that have been ongoing for 50 years (Goulson *et al.*, 2015), should serve as a wake up call for things to come if ecosystem health is not taken seriously or connected to the welfare of humans in general. Ceballos *et al.* (2017) speaks to this point, in an analysis of 27,600 vertebrate species and 177 mammals, concluding: 'Earth is experiencing a huge episode of population declines and extirpations, which will have negative cascading consequences on ecosystem functioning and services vital to sustaining civilization'. Possibly a good sign, CBS, a mainstream news agency, did provide substantive coverage of this research (Gunaratna, 2017). Though many in the 'over-developed' world may not appreciate how they are dependent on other forms of life, more vulnerable humans tend to be 'more reliant on biodiversity and ecosystem services' and therefore suffer disproportionately when ecosystems are degraded and compromised by extractive human practices (Connecting, 2015).

With all this in mind, it is clear that alternatives to what we are currently doing must be considered. The *status quo* or 'business-as-usual' will result in more problems and ones harder to recover from — the more ecosystems destroyed, the less diverse our planet will be, and the more contaminants sprayed/exhausted into the environment, the more toxic it becomes.

1.5. Conclusion

We clearly have many challenges to reaching our goal, **'To provide the highest quality of life to all humans while preserving biodiversity'**. Given human reliance on environmental resources, we recognize that any effort to improve the lives of humans will require a deep understanding and commitment to protecting nature and natural systems as well.

One somewhat obscured theme throughout the discourse so far is the element of time. How long do we have before challenges become intractable? How long do we have to transition? There are two prongs to this. First, we must consider how quickly our current way of life is negatively affecting humans and biodiversity as we know it. Second, we need to ask if any of these changes are irreversible. Regarding the first consideration, we should recognize that the Industrial Revolution is only a few hundred years old and this is but an instant in geological terms. It is also just a short interval of human history as well. Therefore, the changes that we have seen so far are ones that should give us great pause. Regarding irreversibility, we must honestly ask how resilient Earth systems are to change. Current rapid rates of extinctions do not portend well. Precaution should be taken as humanity does not sufficiently understand the complexities in nature and its limits.

One major challenge in achieving any of the above goals is the state of global inequality and immense power associated with private ownership, monetary systems, and political representation. In fact, the power of global elites not only comes from their ability to choose which path they want for themselves but also from their ability to quash or quell the efforts of others who might try to move themselves or others in new directions (with the assistance of the military industrial or the prison industrial complex; see Naomi Klein's *The Shock Doctrine* (2008) and Michelle Alexander's *The New Jim Crow* (2010)). This power imbalance lends itself to autocratic rule and top-down solutions for the benefit of the few at the expense of the many.

It is important to note that about 71% of all greenhouse gas emissions (since 1988) come from only 100 companies (Riley, 2017). Clearly a small corporate elite has made decisions with huge negative impacts as a result of their reckless behavior in the interests of profit generation. However, the '99%' is definitely not in a state of permanent paralysis and given their overwhelming numeric advantage, the future likely lies in their hands.

Another way of understanding the challenge of taking effective action is through analogy. What does one do if one's house is on fire? One gets out of the house as soon as possible, assuming this is possible. However, the suffering of the near billion of chronically-malnourished people is not a 'house' easily escaped from. In 'more developed' countries, intensive poverty serves a similar function through the poverty industrial complex (PIC). Tens of millions of people in the U.S. live paycheck-to-paycheck (or worse); staying alive but in a diminished state that is not easily escaped. This poverty expands when companies that have left for countries where environmental regulations are weaker and where workers are willing to work for lower wages and in harsher, more dangerous conditions — for illustration, consider both the Rana Plaza garment building collapse in Bangladesh (2013) killing over 1,000 people and the Kader Toy Factory in Thailand (1993) killing several hundred and injuring many more. This poverty intensifies when global banking entities rob vulnerable low-income families via sub-prime mortgages, paycheck cashing operations, and high interest credit cards. Historically, it is precisely those that have the least to lose (and the most to gain) that force change to the system (through higher wages, social security, etc.). But they do not do it alone. They often need sympathizers, collaborators, and supporters from the economic and educational elite. As we shall see, the large financing of renewable energy installations (by key financiers as well as sectors of the informed middle class) has been having a huge potential impact on turning the fossil fuel 'ship' back to port where it can be terminated. If a transition to the OWSP is going to be realized, it will certainly take a diverse and sustained effort from a broad range of stakeholders.

We cannot escape the Earth either, no one can — it is clearly the only planet that all of us will live (and die) on. With this realization, we can begin to see how we might act. First, in the following chapters, we look at the opportunities that exist — the means to reach our stated goal. And with these options on the table, we will explore, in the last two chapters, how these options can allow us to reach our goal.

We opened this chapter discussing how catastrophic change may be around the corner, for all of humanity. Yet, we do not know for sure when and how the Earth systems will break down, as tipping points and thresholds are very hard to identify and predict precisely. An objective analysis of the current situation should convince one of the severity of the situation. But with this knowledge, we can dare to consider how to create the OWSP.

Chapter 2
How We Got Here

2.1. The anthropocene

We are now in the Anthropocene, following the Holocene in the geologic timescale. Our genus, Homo, first emerged close to the beginning of the Pleistocene, the epoch of the glacial/interglacial cycles, which predated the Holocene. A concept born in the earth sciences nearly two decades ago (Crutzen and Stoermer, 2000), the Anthropocene is now the buzzword of the 21st Century, not a surprise with tipping points to climate catastrophe and ecological collapse looming ever closer. We are in the epoch of growing human influences on biophysical/biogeochemical cycles, a culmination of anthropogenic impacts dating back to prehistoric time. But many aspects of its intensity have greatly magnified starting with the Industrial Revolution, which gave birth to Fossil Capitalism, and indeed the "Capitalocene" — a renaming which emphasizes the critical role of every expanding capital reproduction and its profound impacts on both humans and nature (Malm and Hornborg, 2014; Malm, 2016). Capital is a social relation between classes in society. Capital reproduction to generate profit as an end in itself has been subject to significant critique (e.g., Marx, 1967; Harvey, 2010; Piketty, 2013). For a very readable exposition of the multifold dimensions of the Anthropocene we recommend Angus (2016).

First, let us go back to the early phase of anthropogenic impact on the Earth's climate. Ruddiman (2003) argued that deforestation some 8000 years ago with its release of carbon dioxide to the atmosphere resulted in climatic warming. Hansen *et al.* (2016) find this hypothesis likely especially since the required anthropogenic carbon emission is lower than Ruddiman assumed. But 8000 years later, driven by the birth and continued reproduction of fossil capital coupled with the scientific technological revolution, the conditions were created for the virtual destruction of a habitable planet for our species, to be sure an unintended outcome, but now a very real potential future. But a more hopeful future is still possible:

> 'Indeed the reproduction of capital, utilizing productive forces powered by fossil fuels, created the material conditions leading to both this bifurcation and its potential resolution, which we now face in the twenty-first century: either the abyss of climate catastrophe driven by carbon emissions, or an ever-diminishing chance of what would appear to be a miraculous escape from this abyss powered by renewable energy technologies. This is the potential negation of the negation (to use the dialectical metaphor) of preindustrial low-efficiency solar power (animal and biomass), through energy derived from fossil fuels, to the high efficiency collection of solar radiation using technologies of wind turbines, photovoltaics, and concentrated solar power in deserts' (Schwartzman, 2016a).

2.1.1. *The industrial revolution*

In the 16th Century England ran out of charcoal because its forests were cut down for wood, especially for its colonial expansion. This led to the growing extraction of coal from its substantial sedimentary deposits. The invention of the steam engine occurred in the early 18th Century as a source of power to pump water out of flooded coal mines, when coal was being mined from below the water table, the surface above the zone of saturation of underground water. Thus, according to this account, the industrial revolution was the deterministic product of technological progress, and as a byproduct the science of modern geology emerged because of the growing understanding of the stratigraphy revealed from mining and canal building driven by demands of commerce. The invention and improvements in the efficiency of the steam engine, especially by James Watt, culminated in the Industrial Revolution — cause and effect.

However, this technological deterministic account has been critiqued, most recently by Zmolek (2014) and Malm (2016), who have emphasized that capitalist relations of production were necessary for the emergence of fossil capital, utilizing steam power driven by coal combustion. Even earlier, the great physical chemist and historian of science J. D. Bernal (1971) partially recognized this point when he wrote:

> 'The Industrial Revolution was closely limited in its place of origin; nearly all its major developments occurred in central and northern Britain...Though the event itself has all the characteristics of an explosive process set off by a particular combination of circumstances that determined the place and time of its occurrence, it remains the final phase of a sustained increase in production over the preceding seventy years or more. Economically this seems to have been determined by the steady growth of a market for manufactured products, mainly textiles, itself largely a consequence of the extended navigations and colonial developments of the seventeenth century' (p. 520).

Zmolek points out that virtually all the technologies, even a primitive steam engine, necessary for an industrial revolution were already invented in antiquity, specifically in the Roman Empire, but such a revolution did not occur because the social relations that were necessary and sufficient had not yet emerged. Zmolek says, '...it was not technology so much as capitalism that was key to the Industrial Revolution' (p. 292). This was the change in social relations necessary and sufficient to make possible the Industrial Revolution.

Similarly, Malm made a profound critique of the technological deterministic birth of the Industrial Revolution by showing that the shift from water power to steam in 19th century Britain was driven by capital's need to control its labor force, not because steam was cheaper than water power. The critical importance of the energy supply is captured:

> '*in the sphere of production*, energy is what makes everything work, and so the control over it will prop up power-as-domination. Indeed, *all* economic activities are ultimately a matter of energy conversion, be they manufacturing, transportation, construction, commerce or drilling, objects in the world can only be transformed, transferred, treated in whatever way by means of energy. At the points of large-scale commodity production,

> that universal force must be concentrated. The power of capital over labour is conditional upon control over it — particularly over its mechanical forms, the ones that set the instruments in motion, without which all production would stand still' (pp. 314–315).

With petroleum joining coal as components of fossil fuel energy supplies in the 20th Century, 'the industrial revolution gave humans the capacity to push energy inputs towards planetary scales and by the end of the 20th Century human energy use had reached a magnitude comparable to the biosphere' (Lenton *et al.*, 2016). By far the dominant capture of energy by the biosphere is by photosynthesis, the conversion of solar into stored chemical energy in the form of carbohydrates.

2.1.2. *Coal, then petroleum, still coal*

Hydrocarbon fuel, petroleum (oil and natural gas) started to replace coal in a significant way in the 1920s when big oil fields in the Middle East were first exploited (Fig. 2.1). Now petroleum dominates the global energy supply, although coal is still much in play. In 2016, petroleum was 57% and coal 28% respectively of global primary energy consumption (BP, 2017).

2.1.3. *Industrial agriculture*

We begin with an extended quotation from Karl Marx (1967), in 1875, *Capital* Vol. 1, describing the consequences of the birth of industrial agriculture emerging with the Industrial Revolution:

> 'Capitalist production, by collecting the population in great centres, and causing an ever-increasing preponderance of town population, on the one hand concentrates the historical motive power of society; on the other hand, it disturbs the circulation of matter between man and the soil, i.e., prevents the return to the soil of its elements consumed by man in the form of food and clothing; it therefore violates the conditions necessary to lasting fertility of the soil. By this action it destroys at the same time the health of the town labourer and the intellectual life of the rural

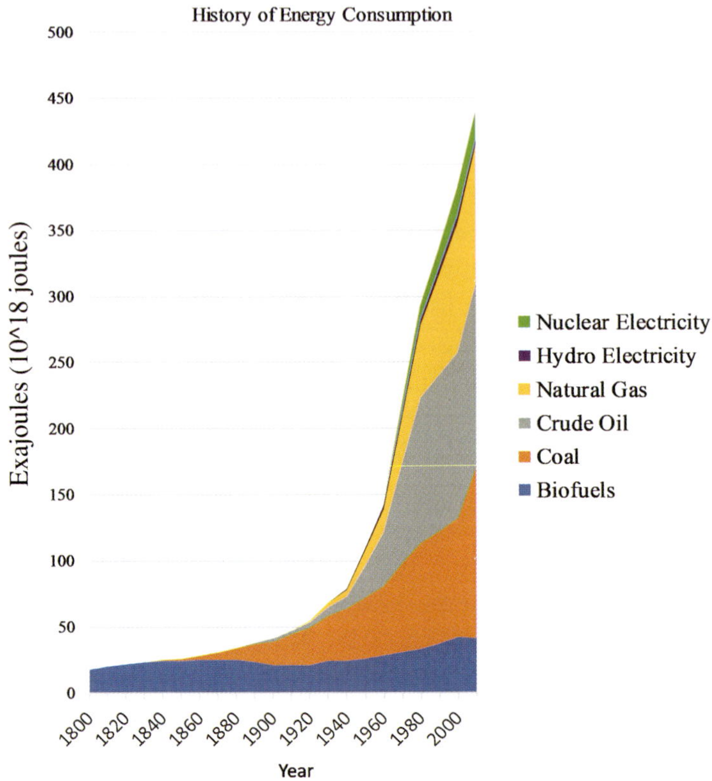

Figure 2.1. History of global energy consumption

Source: The Pennsylvania State University. https://www.e-education.psu.edu/earth104/node/1347.

labourer. But while upsetting the naturally grown conditions for the maintenance of that circulation of matter, it imperiously calls for its restoration as a system, as a regulating law of social production, and under a form appropriate to the full development of the human race. In agriculture as in manufacture, the transformation of production under the sway of capital, means, at the same time, the martyrdom of the producer; the instrument of labour becomes the means of enslaving, exploiting, and impoverishing the labourer; the social combination and organisation of labour-processes is turned into an organised mode of crushing out the workman's individual vitality, freedom, and independence. The

> dispersion of the rural labourers over larger areas breaks their power of resistance while concentration increases that of the town operatives. In modern agriculture, as in the urban industries, the increased productiveness and quantity of the labour set in motion are bought at the cost of laying waste and consuming by disease labour-power itself. Moreover, all progress in capitalistic agriculture is a progress in the art, not only of robbing the labourer, but of robbing the soil; all progress in increasing the fertility of the soil for a given time, is a progress towards ruining the lasting sources of that fertility. The more a country starts its development on the foundation of modern industry, like the United States, for example, the more rapid is this process of destruction. Capitalist production, therefore, develops technology, and the combining together of various processes into a social whole, only by sapping the original sources of all wealth — the soil and the labourer' (pp. 505–506).

It is no coincidence that Commoner (1971) in his seminal book *Closing Circle* on the environmental crisis, with an influence comparable to Rachel Carson's *Silent Spring* (1962), quotes this passage from Marx for its remarkable insight into the roots of the manifold impacts of industrial agriculture. Continuing into the 20th and 21st centuries, the environmental/ecological impacts include the over-fertilization of fresh and coastal marine waters, and the contamination of drinking water with nitrates, pesticides, herbicides, and antibiotics. An illuminating account can be found in Foster and Magdoff (1998). This process will be discussed further in Chapter 3.

The enclosures of common land with transfer of ownership to landlords in the late 16th Century laid the foundation for the expulsion of English farmers that continued into the 19th Century. It led to the demographic transfer to cities and the creation of the industrial working class (Zmolek, 2014; Malm, 2016). The farming commons predated the feudal era, but the 21st Century is the 'true age of enclosure' and the culmination of an ongoing process continuing for hundreds of years (Marzec, 2015). For example, 'Lands with a good potential for development of sugarcane crops, such as Brazil, become ideal sites for enclosure on the pretense of finding ecological alternatives to fuel, such as ethanol production (Marzec, 2015, p. 102).

2.1.4. *The birth of thermodynamics in the context of energy analysis*

The energy supply to society is obviously necessary for the production of material goods, both industrial (now primarily consisting of fossil fuels), agricultural (sunlight for photosynthesis, in addition to fossil fuel used for machinery, processing, transportation as well as the production of fertilizer, herbicides, pesticides — industrial agriculture and GMOs — and of electricity for artificial light sources used in greenhouses/vertical farms), and of course transportation of both people and commodities. Even intellectual/scientific production is possible only by virtue of the supply of energy used for the production of computers, scientific equipment, along with the world-wide dissemination of information both as virtual (the worldwide web, etc.) and as hard copies in the form of books and journals. Critical to an understanding of both the quantitative and qualitative aspects of energy supply to both civilization and nature is the science of thermodynamics, with its focus on heat and temperature and their relation to energy and work.

The science of thermodynamics was a direct result of the practical needs of the birth of merchant followed by industrial capitalism starting in the late 18th/early 19th Century:

> 'The laws of conservation and transformation of energy...were not discovered by disinterested theoreticians motivated by sheer intellectual curiosity; they were the fruit of deliberate attempts to increase the efficiency (and hence the profitability) of steam engines…' (Conner 2005, p. 425).

The experimental determination of the mechanical equivalent of heat led to the formulation of the law of conservation of energy, while the thermodynamic concept of entropy arose directly from Carnot's theorization of the operation of the steam engine (see Bernal, Vol. 2, 1971 and Cardwell, 1989). Wootton (2015) makes the case for experimental science being the source of Newcomen's steam engine, soon to be used to drain coal mines by 1710 (Zmolek, 2014, pp. 323–324). Wootton argues that Papin's understanding of atmospheric pressure and the vacuum, derived from the experiments of von Guericke, Huygens, and Boyle, was necessary for the design of Newcomen's steam engine. Wootton dismisses Marxist historians of science such as Boris Hessen who argued that new science resulted from new social relations, with his prime example being Newtonian physics (Bernal is not even in Wootton's bibliography). Wootton concludes, 'First came the science, then came the technology' (p. 508). But based on the studies of Bernal, Zmolek and Malm already cited, we find the case that social relations have been the necessary and sufficient context for scientific/technological revolutions, such as the Industrial Revolution in the past, with feedbacks both ways between science and technology, persuasive. And as Conner (2005) has shown, the role of farmers, miners, blacksmiths and many others using and improving everyday technologies has been greatly underestimated as generators of the raw material for theoretical science and mathematics.

Further, 'The commodification of social relationships and their subsumption to capital — labour in particular — presupposes the separation of society into a class of property owners on the one hand and a class of "free" wage labourers on the other' (Zmolek, 2014, p. 828). This was the change in social relations necessary and sufficient to make possible the Industrial Revolution.

2.2. The three laws of thermodynamics, not four!

Let us begin with a concise overview of standard thermodynamics and its three laws (see Atkins, 1984, for a clear exposition). The 'zeroth law' regarding temperature is not counted here.[2.1]

The conservation of energy is established by the First Law (and of course after Einstein's famous equation, $E = mc^2$, becomes the conservation of mass and energy, with E being energy, m mass and c the speed of light). The Second Law entails a fundamental property of the universe (but not necessarily the hypothesized multiverse), in which the flow of energy changes in an irreversible manner, measured by the production

[2.1] The Third Law applies to matter at very low temperatures, forbidding it to reach absolute zero in a finite number of steps.

of entropy. There are several different expressions of the Second Law but the simplest is that work can be totally converted into heat but the reverse is impossible. Entropy is defined as the heat supplied to a system divided by its absolute temperature (e.g., 0 degrees Celsius, the freezing point of water, equals 273 degrees Kelvin on the absolute temperature scale), with temperature as a measure of the intensity of thermal vibrations in any material system, its kinetic energy, in contrast to stored potential energy. Zero degrees on the Kelvin scale is the physically unreachable lowest possible temperature, at which, all thermal vibrations cease (the Third Law). Another useful expression of the Second Law is as follows: heat cannot flow from a cooler to a hotter reservoir without any other change, i.e., work must be done. Hence, an increase of entropy is a measure of the increased inability of an *isolated* system to do work, resulting from the degradation of low entropy energy into waste heat. Critical to the definition of an *isolated* system is that it is closed to *both* energy and matter transfers in or out, in contrast to a *closed* system, which is only closed to matter transfers. This distinction between isolated and closed systems is at the root of much confusion, with the most notorious being the fallacious Fourth Law invented by Nicolas Georgescu-Roegen, the founder of ecological economics. Acknowledging both this contribution, while critiquing his Fourth Law, Cleveland and Ruth (1997) observe:

> 'Despite the flaws in Georgescu-Roegen's definition of a Fourth Law, … His focus on the dispersal of materials and limits on recycling foreshadowed the development of industrial metabolism and industrial ecology…in which the analysis of material cycles is used to understand how production and consumption impact the environment, and how to design new technologies that reduce such impacts.'

Historically, the Second Law of thermodynamics was interpreted to imply the eventual 'heat death of the universe'. Whatever the eventual reinterpretation of ultimate heat death, its invocation in the present context and inconceivably far into the future is surely irrelevant to the fate of human civilization in any practical sense (see discussion in Appendix 1).

Unfortunately, Georgescu-Roegen's thermodynamics is still influential, especially among neo-Malthusians and peak oil enthusiasts. His book *The Entropy Law and the Economic Process* (1971) has 6514 citations on Google Scholar as of the end of 2016. In comparison, Commoner's *The Closing Circle*, in our view a more significant contribution of environmental scholarship, had 2413 citations. The continued attraction of Georgescu-Roegen's views is not surprising, but what is disturbing is the appropriation of his fallacious thermodynamic theories to the project of preventing climate catastrophe in a just transition while establishing an ecological bond of society with nature. In particular, if a transition to high efficiency renewable energy supplies is so deemed impossible, the prospects of avoiding climate catastrophe and ending energy poverty now afflicting the majority of humanity are virtually impossible.

Returning to the proposed fourth law, its rebuttal is highly relevant to considerations of whether renewable energy supplies can indeed replace the current infrastructure, dominated by fossil fuels. Georgescu-Roegen (1989) gives two formulations of his fourth law: (1) unavailable matter cannot be recycled; and (2) a closed system (i.e., a system that cannot exchange matter with the environment) cannot perform work indefinitely at a constant rate. Georgescu-Roegen (1981) posits that 'in a closed system available matter continuously and irrevocably dissipates, thus becoming unavailable' and that 'complete recycling is impossible' (pp. 59–60). Here his definition of a closed system follows its standard definition in thermodynamics as already pointed out. If we substitute 'isolated' for 'closed' (an isolated system means there are *neither* matter *nor* energy transfers between the system and its environment) then Georgescu-Roegen's second formulation substituting *isolated* for *closed* is equivalent to the Second Law of thermodynamics. For an economy run on fossil fuel energy, which of course has finite reserves in the Earth's crust, the Second Law simply indicates that energy to do work is not renewable; more on this in 2.4.

The Earth is not an isolated system in a thermodynamic sense because of the incoming solar flux to the surface (and on a sufficient time scale an equivalent radiant energy flux back out to space), but is for all practical purposes closed to matter transfers (except for the trivial fluxes from meteorites landing on and space vehicles leaving Earth). We neglect here the energy flux coming from below the Earth's surface, arising from radioactive decay in the crust and mantle, that is much smaller than the solar flux to the Earth's surface; the geothermal heat flux from the Earth's interior is estimated to be 47 terawatts (Davies and Davies, 2010) — only 0.03% of the incoming 173,000 terawatts of solar flux to the Earth's surface (Archer, 2012). But this geothermal heat flux is 2.6 times the present anthropogenic flux of 18 terawatts (a terawatt is a trillion watts, a power unit, energy per time). However, this energy source from below is the basis of internally generated geologic activity such as volcanism, and is critical to the long term evolution of the crust and biosphere.

Just like the natural biosphere powered by solar energy, the self-organization of the material creation of human activity on the Earth's surface can continue far into the future with the export of an entropic flux into space, provided the long-term energy source of the Sun is utilized.[2.2] It should be noted that the Sun, with its nuclear fusion reactor at its core, is at a safe distance from human civilization, 93 million miles away.

2.3. Ecocatastrophe: the reincarnation of entropy in social prognostication

In his writings, Georgescu-Roegen (see 1971) bridged the gap between entropy's earlier use and the contemporary interpretation bearing on economics, energy and the environment.[2.3] According to Georgescu-Roegen,

[2.2] The Second Law is key to the understanding of the ubiquitous emergence of ordered (so-called anti-entropic) systems in the universe (e.g., stars) and here on earth (e.g., life and society). This spontaneous self-organization of matter is not only consistent with the second law, since entropy always increases in the self-organizing system *plus* its environment, but can also be seen as its very consequence. Ordering and its maintenance within the system generates an entropic flux passing into the local environment (see Bertalanffy, 1968, pp. 40–41 and Prigogine and Stengers, 1984 on the thermodynamics of self-organization). The phenomenon of self-organization and its metatheory is highly relevant to understanding the origin of life and the subsequent evolution of the Earth's biosphere (Sole and Camazine *et al.*, 2001; Bascompte, 2006; and Schwartzman, 2002; 2014a; 2015a).

[2.3] Ironically, Georgescu-Roegen actually leaned at one point (1971; he changes his mind in 1986) to rejecting the heat death scenario because of his favoring the steady-state cosmology (both entropy and matter are created and destroyed) while invoking entropic limits to economic activity, in his critique of neo-classical economic theory ('the ultimate fate of the universe is not the Heat Death... but a much grimmer state — Chaos', neo-classical theory conflicts with the Second Law: the economic process materially consists of a transformation of low entropy into high entropy, i.e., into waste (1971, p. 18) and as low entropy resources run out, especially fossil fuels, economic activity becomes increasingly limited by the accumulation of waste (pollution) and scarcity of energy (for a defense of the orthodox position see Arrow, 1981). Drawing from Georgescu-Roegen's ideas, Daly and Cobb (1989) contend that we are rapidly approaching the physical limits to the further growth of the world economy since the growth of physical throughput will inevitably deplete the energy, materials and space on which it depends, with the concomitant progressive destruction of the biosphere. They further argue that future knowledge cannot remove limits on the physical scale of the economy resulting from finitude, entropy, and ecological dependence.[2.4]

It is important to emphasize that Georgescu-Roegen's 'fourth law of Thermodynamics' is fallacious because it claims that unavailable matter cannot be recycled and that a closed system (i.e., a system that cannot exchange matter with its environment) cannot perform work indefinitely at a constant rate (Georgescu-Roegen, 1989).[2.5] This purported law, as

---

Georgescu-Roegen, 1976, p. 8). Thus, while wavering on accepting the classical Marxist concept of the inexhaustibility of matter in motion on the scale of the universe, Georgescu-Roegen rejects its neo-classical analogue (economic cycle in a finite world without a limit) on the scale of the economy (Schwartzman, 1996, p. 312).

[2.4] This analysis has been challenged (Boucher *et al.*, 1993, referencing Commoner, 1990). See also Sagoff (1995) and Daly (1995) for a recent debate. Daly himself has shown some indications that he has backed off from his original formulation, though it is repeated in the revised edition of Daly and Cobb (Schwartzman, 1996, p. 312).

[2.5] In one paper, Georgescu-Roegen (1976) defines "closed" as entailing "no exchange of matter or energy with [the] environment" (recall that in thermodynamics, such a system is defined as "isolated", not "closed"); he still maintains that according to the second law, matter — along with energy — is subject to irrevocable dissipation (p. 8). This confusion is likely linked to his pessimistic view on harnessing solar energy, since the latter is the relevant energy flux to consider for the closed, but not isolated, system containing economic activity on the Earth's surface. This distinction between closed and isolated systems is also central to the problem of optimizing society's relation to nature. (Schwartzman, 1996, pp. 312–313).

In Jeremy Rifkin's hands (1980; 1989), the entropy concept was extended to its apocryphal limits: entropy as a pollutant, an indicator of cosmic disorder, the inexorable

already discussed, is sheer nonsense since it neglects to account for the possible flow of energy through the system which is defined as closed but not isolated. On the issue of the claim that unavailable matter cannot be recycled, by converting low entropy, high temperature energy (e.g., solar radiation) to high entropy, low temperature heat, work can be produced to recycle indefinitely (see e.g., Bianciardi *et al.*, 1993). Unfortunately, many recent discussions cite favorably Georgescu-Roegen's thermodynamic arguments (e.g., Altvater, 1994; Dryzek, 1994).

While Georgescu-Roegen's views on entropy and the economy are questionable, his work has stimulated welcome and wide-ranging debate on physical-environmental constraints of economic activity and his critique is still relevant for an economy based on non-renewable energy.

---

outcome of all economic activity, and the mother of ecocatastrophe. Note that Georgescu-Roegen enthusiastically endorses Rifkin's treatment of the subject, in his Afterword to Rifkin's book. Yes, Rifkin did support the necessity of shifting to a solar economy, albeit with a strong Luddite flavor, since he favors a pre-industrial global population of less than one billion people (1989 edition, p. 254) and rejects the use of computers since they generate entropy (1989 edition, pp. 190–191). We speculate that Rifkin actually typed his books and papers on a personal computer since it first became widely available in the early 1980s (the senior author, David, got a Mac SE in 1987)! Interestingly, Rifkin has never owned up to his problematic treatment of entropy in his later work (e.g., Rifkin, 2011, 2014)), ignoring the so-called fourth law which he so enthusiastically embraced in the 1980s.

One should distinguish between the entropy of thermodynamics, statistical mechanics, and information theory/computation (see Proops, 1987; Rothman, 1989). The latter two entropies, particularly the statistical mechanical one, have deep, though debatable connections to thermodynamic entropy. The entropy of information theory, especially as a measure of concentration to a set of probabilities (see Proops, 1987), has found wide and useful application in economics and the social sciences. In this discussion, we will only consider the application of classical thermodynamic entropy to economics and the environment. Thus, I will not consider the interesting attempts to apply non-equilibrium and far-from-equilibrium thermodynamics to understand self-organization in the economic and social realms (see Dyke, 1988 and O'Connor, 1991 for examples). A fundamental criticism of Georgescu-Roegen's (and Rifkin's) invocation of entropy is that material/energy transformations in an economy take place far from equilibrium, thus it is incorrect to use the thermodynamic entropy of near equilibrium processes for its description (see Morowitz, 1986). An analogous criticism has been made of its similar use in modeling biotic and climatic processes, but deep insights can be obtained from the near-equilibrium approximation if its limits are appreciated. (Schwartzman, 1996, pp. 313–314).

2.4. Thermodynamic entropy: its use/misuse and redundancy in ecological economics

Does the thermodynamic entropy concept really give us any insight into the environmental effects of economic activity? As a first order conclusion: in an economy run on fossil fuel energy, which of course has finite reserves, the Second Law simply indicates that energy to do work is not renewable (see critique of Rifkin by Rothman, 1989).

Beyond this basic insight, the concept is really redundant to a simple consideration of the energy budget alone in understanding anthropogenic heat pollution and the enhanced greenhouse. Nevertheless, thinking about entropy is useful in gaining insight into issues of recycling, pollution and energy conservation, so in this respect we owe a debt of gratitude for Georgescu-Roegen's provocation.

In addition, the evaluation of alternative ways of accomplishing the same goal (e.g., heating a house or running a car) using Second Law efficiencies (the ratio of the least available work that could have done the job to the actual available work used to do the job) can lead to substantial savings of energy (see Ford *et al.*, 1975; Commoner, 1976). This consideration is especially relevant to global energy requirements in a wind/solar power transition, as Jacobson and Delucchi (2009) and Jacobson *et al.* (2014) have emphasized (further discussed in Chapter 4).

Consider the energy budget at the Earth's surface. A significant fraction of the incoming solar radiation in the form of low entropy visible light reaching the Earth's surface is irreversibly converted into heat radiation (infrared). If the Earth's surface were perfectly reflecting, with an albedo equal to one, then no heat radiation would be emitted from this source. The natural greenhouse effect is caused by the absorption of heat radiation by molecules of water and carbon dioxide in the atmosphere and its re-radiation to the surface. Were it not for this greenhouse effect the Earth's surface would be about 30°C cooler. Any economy based on energy sources other than the direct solar flux impinging on the Earth's surface (i.e., fossil fuels, the stored solar energy of past geological epochs, as well as nuclear and geothermal energy) must inevitably alter the heat budget by the emission of heat radiation over and above the natural flux from the surface. Such direct anthropogenic heat pollution presently accounts for 0.03% of the solar flux impinging on the land surface (Smil, 1992); localized, however, in cities and industrial centers, it produces the heat island effect, the elevation of temperature in and around cities. Much more serious is the well-known, *enhanced* greenhouse effect resulting from anthropogenic carbon dioxide and other gaseous emissions such as methane. And here is a key point: a solar-based world economy would not affect the Earth's surface heat budget, provided the tapping of solar energy involves no net transfers of carbon dioxide, methane or other greenhouse gases to the atmosphere/ocean system (e.g., by deforestation, flooding from big hydropower projects, etc.). Tapping directly into solar energy merely utilizes a small part of the immense flux to do work which ultimately would be simply converted into waste heat anyway, as in the case of the natural heat budget (anthropogenic albedo changes, such as making the surface darker, may result in changes in the surface heat budget, but globally they are small compared to other effects).

Regarding the energy cost of recycling, cleaning up and/or restoring the biosphere, and mining/refining mineral ores with increasingly scarce concentrated sources, the same critical characteristic of the solar energy budget applies, since fossil, nuclear and geothermal fuels all insult the biosphere by incremental heat as well as by pollution effects (e.g., nuclear power results in significant thermal pollution of bodies of water along with the other well-known effects, and has been largely parasitic on fossil fuels). *A non-solar economy must generate additional insults in the cleanup or recycling process*, since its very use must pollute thermally and materially.[2.6]

[2.6] These effects are, of course, on top of the problem of diminishing reserves of fossil fuels and the finite lifetimes of geothermal reservoirs, all which make the renewable energy path imperative. A practical source of fusion energy might eventually be powered by the essentially inexhaustible supply of deuterium in the oceans, with minimal impact on the environment other than incremental heat production, but this option is still far from practical implementation, so a renewable energy transition is imperative. Even if some form of cold fusion should ultimately be developed, the incremental heat problem will remain. A recent estimate of the magnitude of incremental anthropogenic waste heat gives a radiative flux equal to 0.03 Wm^{-2} or 1% of the total radiative forcing from human activities (Chen *et al.*, 2016). However, since the release of anthropogenic heat is regionally localized, the heat island effect is significant, particularly in urban areas.

This is not a necessary outcome for a solar-based economy. It is curious that no recent literature on this subject makes this simple observation in a clearly stated argument that emphasizes its profound significance. Georgescu-Roegen (1976) makes the essential point at least once, but fails to develop it (in later writings, such as his Afterword in Rifkin (1981; 1989), he is less optimistic about the prospects for direct use of solar energy, seeing it as parasites of the current technology (1989, p. 304)). For whether this energy is used or not, its ultimate fate is the same; namely, to become dissipated heat that maintains the thermodynamic equilibrium between the globe and outer space. In a footnote to this passage he points out one necessary qualification: that even the use of solar energy may disturb the climate if the energy is released in another place than where collected. The same is true for a difference in time, but this case is unlikely to have any practical importance (Georgescu-Roegen, 1976, p. 28).

2.5. More on recycling

Paul Burkett is a well-known Marxist scholar with many valuable publications (e.g., Burkett, 2003; Foster and Burkett, 2004; Burkett's 2005 paper is included with minor revisions as chapter five in Burkett 2006; Foster and Burkett, 2016). Burkett (2005) supported Georgescu-Roegen's theory of entropy in an attempt to seek convergence of Marxist theory with ecological economics. Unfortunately, the very shaky foundations of Georgescu-Roegen's thermodynamic theory, however, undermine this attempt.[2.7]

[2.7] More specifically on the possibility of 'complete' recycling in an open system, Burkett's discussion of this issue (Burkett, 2005, p. 132) lacks sufficient concreteness with respect to a real physical economy on the Earth's surface, consistent with Georgescu-Roegen and Daly's abstract treatment. In practical terms, 100% recycling efficiency is not required (see Kaberger and Mansson's (2001) illuminating discussion). Given the possibilities of a future dematerialized solar economy, with a lower throughput than now, and of course recognizing that current information technology is not really dematerialized under current capital reproduction, as Burkett (2005,135) rightfully argues, the huge solar flux is again the basis of any ultimate limit to practical recycling on the Earth's surface, and not the entropic flux of waste heat. The latter would be dissipated anyway by the absorption of solar energy on a land surface (with an albedo, i.e., reflectivity, of about 0.3–0.4, with 0

We will now make a personal observation to hopefully illuminate why recycling reduces energy use. The senior author's now deceased father (and the junior author's grandfather) spent 40 years as a diamond setter on the Bowery in lower Manhattan, in partnership with his two older brothers. He collected the filings of platinum and gold in a metal tray below his workspace. He and his brothers then sold the filings to be remelted. The alternative was to simply throw these filings in the trash can, ultimately ending up dispersed in a landfill, still potentially recoverable but only with the expenditure of significantly more energy than what was the actual practice. Thus, the energy gain in recycling is

---

being perfectly absorbing and 1 being perfectly reflecting (like an ideal white surface). Under these conditions, the 'tremendous increase in the entropy of the environment' or the 'adverse material effects of waste heat on eco-systems' resulting from recycling (Burkett, 2005, pp. 132–133) is an illusion for a solarized economy as Kaberger and Mansson (2001) show. Unfortunately, Burkett's critique of the case made for the plausibility of total recycling in an industrial society (citing Ayres, 1999) does not confront the qualitative difference between a solarized and a depletable-energy-based economy. Further confusion is found in Burkett's quotation from Georgescu-Roegen: 'At the macro-level no practical procedure exists for converting energy into matter or matter of *whatever form* to energy'. It is not clear what 'whatever form' means. In a footnote, Burkett cites Daly (Burkett, 2005, p. 120; Daly, 1992) in support. In this reference, Daly says: 'Although we can turn matter into energy, we have no means for turning energy into matter on a significant scale'. Daly is clearly referring to nuclear reactions, where mass to energy conversion is small but measurable, unlike chemical reactions where the conversion likewise occurs but is infinitesimal. Burkett critiques energy reductionism in his citation of Georgescu-Roegen (Burkett, 2005, e.g., p. 121, Footnote 14). Is it energy reductionism to uphold the relevancy of the Second Law, i.e., entropy must be considered besides energy, entropy in its full quantitative and qualitative aspects (see discussion of the entropy of mixing and its relevancy to recycling and pollution in Schwartzman, 1996)? Ignoring the Second Law is indeed energy reductionism. The issue of friction and dispersal of matter in anthropogenic cycles has energetic, biogeochemical and social qualitative aspects which some critics of Georgescu-Roegen take seriously, but that does not make the 'fourth law' any more valid. Friction equals waste heat; dispersal of matter can be radically reduced depending on the physical design of the process of production/consumption and, of course, energy source. Two of Georgescu-Roegen's examples of 'unavailable matter' arising from the inevitable friction inherent in any physical process are rust and broken glass (Georgescu-Roegen, 1986). So, we are to believe that even with available energy these wastes cannot be efficiently turned back into iron and glass bottles respectively?' (Schwartzman, 2008, pp. 54–55).

demonstrated. Industrial design and environmental policy are critical aspects of the efficiency and energy requirements of recycling and waste production.

The study of natural ecosystems gives more insights into the recycling process. The senior author's collaborator of long standing, Tyler Volk, defined the cycling ratio as the ratio between the flow within an ecological cycle and the flow into/out of this cycle at steady-state (Volk, 1998). This concept can provide insight into the potential efficiencies of future solar industrial production/consumption. For example, the very high cycling ratios achieved by ecosystems for several elements (e.g., potassium, essential to life) with relatively small fluxes into the biosphere suggest that natural systems are useful models for industrial ecology and a sustainable future (e.g., Ho and Ulanowicz, 2005). Therefore, we welcome a long overdue collaboration between ecological Marxists, ecological economists, ecologists and biogeochemists, informed not only by the cutting edge of science and technology, but also by the cumulative wisdom of actual practice by indigenous people around the World. This cross-fertilization of disciplines is imperative to confront the challenges of the 21st Century.

In conclusion, we have now examined how we got into the current historically-unprecedented crisis facing humanity and biodiversity as we know it, followed by an in-depth discussion of thermodynamics, particularly the importance of a clear understanding of its Second Law with respect to the energy question so critical to our climate change challenge: the transformation of agriculture from the present mode, dependent as it is on fossil fuel energy. With this as a foundation, in the next chapter, we will look in more depth at the current challenges regarding energy and food.

Chapter 3

The Centrality of Energy and Food

3.1. Introduction

As introduced in Chapter 1, there are many challenges faced by humanity as we conclude the first score years of this century. If we could tackle/solve only one problem, which problem should it be? This is a question that often gets asked of environmentally-minded people. It is a question that forward thinkers in other areas ponder as well. And it should be a central question that we all give considerable attention given where we are and the limited time we have to 'right this ship'. While it presupposes, probably wrongly, that there is one problem that, if solved, would lead directly to the solution of many other problems, it does focus us on developing a deeper understanding of the interconnectedness of all our challenges as well as prioritizing among them.

Our analysis of global environmental challenges (along with other such analyses, by organizations/institutes such as the Worldwatch Institute) leads us to believe that there are two major areas that hold priority status as it pertains to our future efforts and actions. If we are able to provide each human being with sufficient energy and highly nutritious food, and do so in ways that are in harmony with natural systems, humanity will have entered into a new era, one that will greatly improve the quality of life of humans, with the planet left habitable for biodiversity at levels close to the present. These two components of human civilization (energy and food) are not only primary to our day-to-day existence; their inherent interconnectedness to other necessary resources (such as water, building materials, etc.) makes them special in scope and relevance and, therefore, deserving of our immediate and undistracted attention.

Conventional B.A.U. (business-as-usual) projections tell this story: as human populations continue to grow with no plateau expected until the next century (United Nations, 2017) and levels of affluence continue to rise, we are clearly going to have to produce more food, energy, and 'stuff'. Further, these projections claim that even when human populations stabilize, affluence levels will likely continue to rise. Hence, we submit that changes in demographic as well as consumption patterns must materialize — these will be explored in future chapters. We argue that two key resources, energy and food, will be the key determinants of future health, prosperity (radically redefined) and planetary recovery. These two resources are not only critical because they are (and will be) required in great quantities, but the means by which we obtain them, disseminate them, utilize them, and discard them all have tremendous impacts on every living thing on *terra firma* (the land) and in *terra agua* (the sea). Currently, not only is a huge percentage of our financial capital spent on them, but the dominant modes of obtaining them are done in ways that are not sustainable. Furthermore, these two resources are in many ways drivers of consumption of other resources as well. Thus, making better decisions about how we obtain and use them will influence almost all other resource demands. Sound choices in these areas will translate (though not guarantee) wise choices in other consequential areas.

While a few billion people on Earth have sufficient access to food, water, and energy, many more than that go without one or more of these resources on a regular basis. If nearly a billion people on Earth suffer from chronic malnutrition (the number has varied from 825 million to 1.02 billion since the 1970s; Nierenberg, 2013), 'Houston, we've got a problem'. Obviously, every effort imaginable should be made to make sure that this challenge is effectively confronted. A comparable but less recognized evil is the energy poverty that exists in the world today. Without basic allotments of energy, many people around the world cannot satisfy basic needs, such as, cooking food, heating/cooling their homes, or performing important tasks at night; consider that 1.4 billion people do not have access to electricity (OECD, 2010). Even in places where some energy is available for such activities, it is often dangerous (e.g., kerosene) or detrimental to the local environments and health (e.g., burning firewood). Without sufficient food or energy, more than a billion people suffer unduly, and when water limitations are added to the mix, the number detrimentally impacted grows even further. Importantly, the ramifications of having so many people literally deprived of at least one essential resource on global peace and tranquility cannot be overlooked either.

3.2. Major resources necessary for production of food

Food requires massive amounts of resources to produce and provide. Until early forms of agriculture emerged, humans obtained much of their food directly from the land and sea around them, in the form of native plants and catchable animals. This hunting and gathering required human energy and some resources, in terms of hunting equipment, cookware and storage containers. However, modern agriculture demonstrates how far we have come in terms of the resources required to feed humans today. Consider, the food on one's plate in a 'developed' world context. Most of the ingredients that make up a meal were likely grown, harvested, processed (multiple times), packaged, transported (thousands of miles, perhaps more than once), refrigerated (or dosed with industrially-formed preservatives), before finally being cooked and arriving on your table. Clearly, each increase in steps as well as the complexity of the process come at a higher energy cost. However, the resource most fundamental to our modern food system is land itself.

The past 300 years has seen many changes but none have been any more impressive than the tremendous expansion of land for agricultural purposes. Currently, crops occupy about 1.6 billion hectares (ha) while pastures take up an additional (roughly) 3 billion ha (in 2000; Lambin and Meyfroidt, 2011). These combined amount to some 40% of the Earth's land area — up from 7% in the 1700s (Owen, 2005). This is a significant number given its contrast to the 95% of humans who currently live on only 10% of the land (European Commission, 2008). A comparison of these percentages clarifies that the claim, 'there may not be enough room for more humans (on Earth)', has a lot more to do with the land required by agriculture than the land occupied by humans themselves — by the way, we think this claim is misleading, something we will explicate in

Chapter 6. The "Green Revolution" in agriculture negatively affected nearly 3 billion people by degrading their land through unsustainable short-term yield maximization (Nierenberg, 2013). By 2030, it is anticipated that cropland acreage will increase another 7% and pasture needs will increase by up to 5% (Lambin and Meyfroidt, 2011). These increased allotments are likely to put further strains on wildlife and reduce biodiversity.

Clearly, no other activity takes up more land than human efforts to feed ourselves. However, is there enough land to provide humans, now and in the future, with sufficient calories and nutrients? Unequivocally, the answer is 'yes'. Current hunger and malnutrition are not the result of insufficient land or overpopulation (Biel, 2017), rather they are the outcomes of maldistribution, profiteering, and massive (unnecessary) amounts of food waste (Angus and Butler, 2011). In the U.S., this waste (over 1,200 Calories per person — here, a Calorie is equivalent to a kilocalorie) has reached epic proportions (Barclay, 2014). On all continents, *per capita* food supply currently exceeds 2,600 Calories with the global average being approximately 2,900 Calories — a great increase since 1961 when the global average was about 2,200 Calories (Roser and Ritchie, 2017a). This caloric supply should be more than sufficient to provide all humans with food Calories since the National Health Service in the United Kingdom asserts that an average male adult needs only about 2,500 Calories per day to maintain weight (and, consequently, the average human need is quite a deal less, when children, the elderly, and women are factored in). Additionally, many millions of acres are currently being utilized to make biofuel, particularly ethanol — nearly 40% of corn grown today in the U.S. is used for this purpose (Nierenberg, 2013). Also, the majority of acres of agricultural land in the U.S. is used to feed carnivorous diets, many in excess. Regarding the future, as we will show in Chapter 5, agroecology has the capacity to maintain these caloric levels without the harm done in the name of modern industrial agriculture. Sufficient land for food is not and should not become a problem. Yet, the impacts that this use has should not (and cannot afford to be) ignored.

It goes without saying that agriculture also uses lots of water. However, industrial forms of agriculture are legendary for overusing water resources. At the moment, about 69% of the world's freshwater withdrawals are done for agricultural purposes (Global Agriculture, 2017) though this percentage excludes from the total a slightly larger flux that is extracted for thermoelectric electrical production (USGS, 2010). Many crops currently grown occupy land where the natural supply of precipitation is insufficient for reasonable harvests. Much of California, now the largest producer of vegetables in the U.S., owes its abundance to water extracted from other areas (as documented in Marc Reisner's classic book, *Cadillac Desert*, 1986), not based on abundant sources of natural precipitation; irrigation for food (crops and meat) currently accounts for 77% of freshwater withdrawals in California (LaFond, 2015) (once again, ignoring thermoelectric extractions which is nearly 1.5 times larger; CGS West, 2005). Much of current farming of wheat, corn and soybeans in the Central Plains can only be sustained with the use of the Ogallala aquifer, one of the largest underground sources in the world which sits below ground in eight U.S. states. In fact, 90% of the irrigation derived from this mighty aquifer is used to meet agricultural needs in the region (Little, 2009). Since 1950, about 9% of the Ogallala has been depleted and at current rates of extraction, the aquifer could be completely depleted in just a few more decades (Amelinckx, 2015). Worldwide, the story is worse for many — particularly in parts of Africa where natural water allotments are not sufficient even for everyday human activities, and far from sufficient for expanding local supplies for the increased agricultural activities necessary to meet a growing population's needs in the future (Klare, 2002). China, home to the greatest number of humans on our planet (nearly 20%) as of 2017, contains 21% of the World's total irrigated land but less than 7% of the renewable freshwater (Postel, 2013). Northern India, home to most of the (currently) second most-populated nation's agriculturally rich land has seen staggering levels of groundwater depletion in recent years — a problem deemed considerably graver by the fact that at least 15% of food grown in India is supported by these mined resources (Postel, 2013).

To give an idea of just how much water is required to produce some of the common foods that we eat, consider these unexpectedly high numbers. Mekonnen and Hoekstra (2010a and 2010b) did an extensive international study on the diets and water consumption of farm animals, as well as the water requirements for plant crops. They conclude that producing animal mass (on a per 1,000 kg basis) requires this amount of water: beef cattle → 7,480 m^3; sheep → 4,520 m^3; pigs → 3,830 m^3; goats → 3,080 m^3; and, chicken (broiler) → 3,360 m^3 (Mekonnen and Hoekstra, 2010b). Considering that 1,000 gallons is ~3.8 m^3 (and 1 gallon of water weighs ~8.4 lbs.), beef equates to 7,520 lbs. of water for each pound of meat, while the other meats (per pound) require roughly 4,540 lbs. for lamb, 3,850 lbs. for pork, 3,090 lbs. for goat, and 3,380 lbs. for chicken (broiler); note, here "meat" refers to all parts of the animal, including those that might not normally be eaten by humans. Crops generally don't require as much water as meat, but the numbers are still shockingly high, see Table 3.1. Obviously, a typical Western meal (~1–2 lbs.) requires

Table 3.1. Water used in production of various foods

Food (1 1b, if not otherwise specified)	Water used in production (lbs)
Apple	700 (2)
Banana	860 (2)
Bread	1,612 (1)
Cabbage	200 (2)
Cheese	5,000 (2)
Chocolate	24,000 (2)
Coffee (1 cup)	140 (1)
Corn	919 (1)
Egg (1 egg)	418 (1)
Lettuce	130 (2)
Mango	1,600 (2)
Olives	4,400 (2)
Orange	460 (2)
Potato	250 (2)
Rice	3,400 (2)
Sugar (cane)	1,500 (2)
Tomato	180 (2)

Sources: Adapted from (1) USGS (2017); (2) Hoekstra (2008).

several thousand gallons of water — much more than a pot of boiling water or the volume of the refreshment used to 'wash down' the food. And again, we can see the implications of diets with significant meat consumption versus those without. And it is not only food agriculture that requires a lot of water. A cotton t-shirt requires 2,500 liters of water to produce, most of which is used to grow the cotton itself (Postel, 2013).

In addition to the overreliance of current agriculture on exhaustible resources, climate change may present greater challenges by redistributing precipitation. Such change is underway — it is not just an imagined future reality — and this has resulted in the need for more irrigation installations. Marshall and Aillery (2015) point out that 'groundwater supplies are in decline across major irrigated areas, and surface-water supplies for agriculture are particularly vulnerable to shifts in precipitation, water cycling, and demand for water in nonagricultural sectors'. Furthermore, in the United States, declines in irrigation over the course of the century as a result of aquifer depletion or expected increases in precipitation, both make future required investments problematic (Marshall and Aillery, 2015). Climate change is expected to result in declines in most crops in the United States. Marshall and Aillery (2015) anticipate 9–15% declines in key crops such as corn, soybeans and oats in 2020 and 2040, with additional larger declines as climate change intensifies further into the 21st Century. Other regions of the world where available water resources are currently very limited face even greater tragedies if climate change shifts this precious resource to other venues.

For the bulk of humanity's existence, agricultural labor was done with human muscle and, then, eventually it was augmented with animal power. Modern food production, however, particularly with grains and increasingly with other crops, relies heavily on fossil fuels to create energy to do work. In 2007, the energy use in the U.S. food system was some 16 quadrillion (10^{15}) Btu, which amounts to 16% of total U.S. energy consumption (Azzam, 2012). Fossil fuels run the large machines on the farm, such as tractors and combines, and in the processing plant they enable the drying (and possible roasting) of the grain and seeds. To illustrate the heavy resource demand of industrial agriculture, consider that it currently uses 70% of the resources consumed in agriculture, but only produces 30% of the food produced. In contrast, only 30% of the resources are needed by the 70% of food that is still produced by family farmers (Brescia, 2017). Additionally, and less obviously, fossil fuel-based fertilizers, synthetic nitrates in particular, are obtained through the reaction of natural gas and steam, which removes the oxygen molecule from the chemical composition and leaves ammonia (once CO_2 is removed). Similarly, fossil fuel is used in the preparation of pesticides, and antibiotics and hormones (for animals).

However, interestingly, the highest percentage of the energy consumption associated with food is not used on the farm (which accounts to about 14% of this total, in 2002) or to make fertilizers. Rather, energy is consumed largely in the commercial (wholesale/retail) and at-home components, each of which accounts for about 28% of the total (Canning, 2010). Surprisingly, the proportion of energy spent on processing (19%) is also larger than that of production alone. Also, the average growth rates in commercial and at-home energy use remain positive, despite the incorporation of energy-saving technologies, while energy use on farms and in packaging is now declining (Azzam, 2012). This suggests that efforts to reduce waste, which is often solely focused on farming practices (under the guise of 'efficiency'), needs to focus as well on what is going on at the marketplace (handling and processing inputs) and in the home (transportation to-and-from homes, cooking methods, refrigeration practices, food waste, etc.). Fortunately, all of these concerns are now addressed in the UN's *Sustainable Energy For All Initiative* (SEFA, 2017). Obviously there remains a great deal of opportunity to be much more efficient with our agriculture practices at all stages of production and dissemination. However, as we will see soon, efficiency isn't necessary the best guide for assessing changes in modern agriculture.

As mentioned, industrialized agriculture depends on the use of large machinery, which currently runs almost entirely on fossil fuels. The efficiency with which food is produced in the non-renewable energy-dependent conventional agriculture system is decreasing, as observed in the fossil fuel energy used per food energy produced now equaling 9:1 (in the 2000s) — higher than the 7:1 ratio observed in the 1950s. Additionally, despite 'the widespread application of crop breeding programs' within conventional intensive agriculture, yields have failed to improve for several major crops, including cereals, which have stagnated since 2000, in part because of degraded soils that have been treated solely as 'a physical substrate to support plants' (Cameron *et al.*, 2015). Engel-DiMauro's (2014) book, *Ecology, Soils, and the Left: An Eco-Social Approach*, puts the challenge of soil erosion/degradation in a historical, political economic, ecological context, precisely what is needed to transition to global agroecologies (see Chapter 5).

One key aspect of conventional agriculture's attempt to increase yields (per acre) has been the widespread use of synthetic chemicals to provide nutrients to soils or to eliminate pests. Synthetic fertilizer has grown in use since the Haber process — a means to fix N_2 (g) (from the air) into nitrogen that can be utilized by plants, i.e., in the chemical form of nitrates — was established in 1909. Haber's process wasn't just lucrative for farmers but also as a means to create explosives that were desired in the two World Wars that were to follow (Philpott, 2013). Post-World War II and the advent of nuclear weapons, nitrate plants built to make bombs found a new customer: the fields of the planet — particularly large swaths of grain and, in particular, those growing corn in the United States. Since then, large quantities of natural gas have been utilized in this way and this activity currently consumes 5% of the world's natural gas (and about 2% of the World's energy supply) (USDA, 2017). In the U.S., commercial fertilizer consumption peaked in 1981 at 24 million tonnes (MT), but has remained above 20 MT in all but six years since (USDA, 2017). Approximately 30% of food's fossil fuel use within the U.S. goes into making chemical fertilizers (Oliver, 2008). Nitrogen fertilizer use in the U.S. is ~13 million tons (MT), about half of which is imported (in 2011; USDA, 2017) and about 47% of which is used on corn crops (Philpott, 2013). Phosphorus (chemical symbol, P; usually in phosphate form, another common macronutrient in industrial fertilizer mixes) must be mined from non-renewable stocks (phosphate rock) that are diminishing as well (Cameron *et al.*, 2015). Potassium (chemical symbol K, usually cited as potash, K_2O — found in various mined and manufactured salts) represents another major macronutrient used throughout the world. In the U.S., phosphate and potash use is about 4 MT per year, remarkably constant for the past 50 years (USDA 2017). About 30% and 80% of phosphorus and potash, respectively, used in the U.S. is imported (in 2011; USDA 2017). See Roser and Ritchie (2017a) for a multitude of graphs of worldwide fertilizer, nutrient and pesticide use over the past century.

Globally, NPK (a common way these three common macronutrients are cited) was applied in fertilizers only marginally prior to 1940 (<10 MT total; Steffen *et al.*, 2015), but the period post-World War II saw dramatic increases. Annual consumption is still growing from 162 MT (in 2008) to 183 MT (in 2013) — a 13% growth in 5 years — and is expected to grow to 201 MT in 2018 (another 10% increase in the subsequent 5 years) (FAO, 2015a). In the U.S., there has been a shift away from multiple-nutrient fertilizers allowing 'farmers to apply precise amounts of a specific nutrient for plant use at the least cost' (USDA, 2017). For comparison, 201 MT is equivalent to 65 pounds (or 29.5 kg) per year per person on Earth. Additionally, the U.S. consumption figure is ~11% of the total, despite being home to less than 5% of the world's population. Globally, the annual growth in NPK use is ~2%, but Africa, East Europe and Central Asia have growth rates near double this and are anticipated to continue to do so in the near future (FAO, 2015a).

Pesticides are also largely created using fossil fuels, as the chemical compounds constituting them are derived from actual carbon bonds found in fossil fuel molecules. Pesticides have also seen a great surge in demand in the 20th Century. Pesticides are used to kill "pests", mainly plants, insects, and nematodes (although rodents and fungi come into play sometimes as well). Fungicides make up <7% of all pesticides used in the U.S. since 1971 (Fernandez-Conejo, 2014). The use of pesticides has grown since World War II, similar to the case for synthetic fertilizers, driven by the shift from supplying the military with poisonous chemicals to using them for production of pesticides to be applied on increasingly large acreages of crops. Huge expansions in use have occurred globally during the past 70 years, from some 100,000 tonnes in 1945 to an estimated 3.8 MT in 2000 — a 38-fold increase (Tilman *et al.*, 2002). This latter value amounts to about 1.4 pounds per human in production annually (notably, the USEPA estimates smaller world usage at ~2.4 MT in 2007; EPA, 2007). Usage in the U.S. is still very sizeable as well despite the increase in GMO crops. For 21 select crops, usage grew from 196 million pounds (of active ingredients) in 1960 to 632 million pounds in 1981, to 1.13 billion pounds by 2006 — amounting to nearly 4 pounds per U.S. resident, with herbicides making up about 45% of the total weight (EPA 2007). Declines have been observed more recently (e.g., in 2008, usage was 516 million pounds; Fernandez-Conejo, 2014) though data is actually hard to come by. For instance, data is available only as recently as 2006/2007 from the USEPA — apparently when George W. Bush was president, the EPA stopped issuing pesticide use data every two years, something they had agreed to do in the past. However, FAO does have some international data (though, interestingly, no data is available from Russia, Nigeria or Indonesia, some of the world's most populated countries) and it shows the following recent change in pesticides use: China (+41% from 2000–2014), Brazil (+180% from 2000–2013), USA (–7% from 2000–2007) and India (–2% from 2003–2010) (FAO, 2014). Notably the small declines in India are part of a shift to bio-pesticides nationwide, though many banned pesticides are still produced in India and sold to export markets (Sood, 2013). Hormones and antibiotics provided to animals also require a significant amount of energy during production and this energy comes primarily from fossil fuels as well.

Energy also has a giant footprint on our planet. Not only does energy require a lot of land and water but it also demands tremendous resources to secure. Regarding land, all forms of energy require it. Land is utilized in the extraction, distribution, refinement, dissemination, and disposal of energy (and its byproducts). Renewable energies require significant amounts of land to secure adequate amounts of the resource, although solar panels can often be put on existing roadsides and rooftops (more on this in Chapter 4). However, Allred *et al.* (2015) conclude that oil and gas extraction has far reaching and long lasting negative impacts on life. Their work represents the first to substantively look at modern oil and gas extraction 'combining high resolution satellite data of vegetation dynamics with industry data and publicly available data of historical and present-day oil and gas well locations' in much of central North America. They estimate that due to the 50,000 new wells drills each year since 2000, biomass loss (measured in terms of net primary productivity (NPP)) has been about 4.5 Tg of carbon or 10 Tg of dry biomass, and that the land area occupied by 'well pads, roads, and storage facilities built from 2000 to 2012 is 3 million ha, the equivalent land area of three Yellowstone National Parks' (Allred *et al.*, 2015). Clearly, this timely work calls into question reports/studies which have claimed how small the land impacts of fossil fuel activity are. We will compare this land area to that needed for a 100 percent renewable energy transition in Chapter 4.

The substantive land use required for energy today (as well as the tremendous economic costs that are borne by it — to be discussed in a moment) suggests that waste should be a major concern. Energy efficiencies of various energy sources has been a topic of great attention as it relates to their consequent physical and financial waste. For the past century, efforts have been made to make fossil fuel technologies more efficient in converting the potential energy of the carbon bonds into forms of usable energy. Different ways of burning coal or oil result in different amounts of usable energy. Renewable energy has also focused heavily on efficiency. Solar panels convert a given amount of incoming (shortwave) sunlight. Hydropower installations convert a given amount of water into mechanical energy using gravity to compel the water to move to a lower potential. Despite all the effort, many fossil fuels are still largely inefficient, in particular, if electricity is the desired energy output. Coal fired (electricity) and nuclear power plants run at about 33% efficiency while natural gas plants run at about 43% efficiency (US EIA, 2015). Automobiles, huge consumers of gasoline worldwide, typically have efficiencies between 15–37%; but when considering that the goal is to move the passenger(s) and not the vehicle itself, these efficiencies drop to less than 5%. Light bulb efficiency is typically 3% with incandescents but has improved substantially with LEDs to the range of 20–30%, not to mention that LEDs last ten times longer as well. Obviously, when more than half of energy's potential is lost (typically as excess heat that has no immediate use), alternatives deserve a closer look (more on this issue in Chapter 4).

However, before continuing, a critical point must be made. All of these efficiency numbers belie the true efficiency of the process at hand. A major purpose of burning coal is to provide electricity to a community. A true efficiency should be inferred by considering all the inputs of energy necessary to obtain the kWh at the end. These inputs include literally hundreds of steps involved in extracting the coal from the mountain to burning it in the furnace. Scientists attempt to account for all this energy by using the term, embodied energy. Comparing the embodied energy of various energy sources, captured by the Energy Return over Energy Invested Ratio (EROEI), will be a subject of the next chapter. Suffice it to say here that efficiencies as they are commonly used in most comparisons of energy types fail to account for this much more holistic and valuable term.

Some forms of energy require a great deal of water as well. Nuclear power plants need very large volumes of water to keep them cool — they pump in water from rivers or lakes and pump it back out warmer than it was on intake. This thermal pollution does detrimentally impact local ecosystems. Coal and natural gas fired power plants also use great amounts of water and this is why you find them located near large water sources. It is estimated that in the U.S., 41% of all freshwater withdrawals (and 49% of total water use) are for this purpose (USGS, 2005). Recent droughts in the U.S. have actually caused a temporary shutdown of a nuclear power plant (in New London, Connecticut) and the contemplation of closing many others. Water withdrawals in the U.S. have been steady at around 200 billion gallons per day from 1985–2005 after showing a steep increase from 50 billion gallons per day in 1950. Globally, such increases are expected as fossil fuel energy 'development' continues.

Refining petroleum also requires a great deal of water. Current efforts to do this in the U.S. amounts to 1–2 billion gallons of water used each day (McMahon and Price, 2013). And efforts to create ethanol as a 'green energy fuel' come at a high cost of water as well, as they use 3–100 times more than that used in refining petroleum (Wu and Chiu, 2011).

3.3. Industrial agriculture and fossil fuels are hazardous to our future

Dominant systems of energy and agriculture not only require great amounts of resources, they also take a tremendous toll on humanity and nature. Virtually every aspect of both systems produces waste streams damaging to life forms, including humans. The hazardous waste streams associated with their components often begin as deliberate inputs — such as, pesticides, hormones, antibiotics and fertilizers in the case of agriculture, and fracking liquids and cyanides in the case of natural gas and oil. In other cases, naturally occurring trace elements, such as mercury in coal, are toxic in of themselves. The combustion of fossil fuels (and biomass) produces secondary (and tertiary) chemicals, some of which are poorly understood and others of which are known carcinogens. The roughly

100,000 synthetically produced chemicals in our environment, in part produced to enhance crop production, reduce crop losses, and maximize energy extraction rates, become part of the Earth's many reservoirs. This toxic soup results in countless numbers of secondary chemicals being formed, many never occurring naturally until their synthetic precursors were first introduced into the biosphere by modern industrial technologies. Additionally, these toxic chemicals not only impact those who work closely with (or near) them, they get moved from the warmer regions of the planet to the colder ones, often far away from large metropolitan areas (Sadler and Connell, 2012). The meteorological process that directionally redeposits these toxic compounds can have demonstrable impact on indigenous communities — ones obviously not responsible for the toxins themselves. Case in point, breast milk among Inuit mothers appears to have the highest concentration of dioxin among world populations (Johansen, 2002) and the concentration of PCBs in their milk is 5–10x higher than that of mothers in Southern Canada (Nierenberg, 2013).

Most research on the human effects of industrial chemicals focus on the impacts of long- and short-term exposure to individual primary chemicals. Even these tests are greatly under-researched with only about 10% of these chemicals having been tested for effects on human health (Salter, 2013; Urbina, 2013). The results of these studies establish that many widely used chemicals (such as atrazine and Roundup) are dangerous for a variety of reasons. They are carcinogenic, teratogenic, immune disrupters, and/or neurologically damaging (Steingraber, 1997). Organochlorines, creosote and sulfallate are well-known carcinogens, and others like DDT, chlordane and lindane are tumor promoters (Dich *et al.*, 1997). According to the World Health Organization, the most widely used herbicide, glyphosate, and a few other commonly used ones (malathion and diazinon) were deemed 'probably carcinogenic to humans', and two other pesticides, tetrachlorvinphos (TCVP) and parathion, are 'possibly carcinogenic to humans' (Cressey, 2015). It should be noted that the WHO very recently changed its assessment on glyphosate (see later discussion). Noteworthy, several agricultural chemicals used widely in the U.S. have been banned in Europe because of their suspected health compromising characteristics; the Stockholm Convention is now the basis to identify new chemicals for elimination and the U.S. does little in response. Almost no studies are done on the impacts of simultaneous exposure to multiple primary chemicals, as this is very difficult to do given the vast number of chemicals to study (Shukla (2017) reports that more than 12 pesticides were found on individual samples of strawberries, grapes, snap peas, cherry tomatoes and bell peppers produced by contemporary industrial farming practices) and their combinations are, therefore, virtually innumerable. Additionally, very few studies look at the effects on the most vulnerable of us, namely children, elderly, and people with compromised immune systems (Steingraber, 2001). Typically, epidemiological studies assume that the exposed person is a healthy adult male (Steingraber, 1997). Additionally, and profoundly, very, very few studies test the effects of the secondary chemicals that are formed when the toxic soup undergoes reactions (when exposed to each other under varying environmental conditions such as heat/cold, frost/thaw, precipitation, ultraviolet light, etc.). Hence, we are definitely subject to a game of 'Russian roulette' with human (and ecosystem) health each and every day as we continue to be exposed to mass produced/consumed primary synthesized chemicals. And since many of these chemicals are persistent in the environment, lasting several years before breaking down, the untold damage they wreak will continue into the foreseeable future. Additionally, when these chemicals do break down their new forms are not necessarily innocuous either.

The extraction and processing of fossil fuels can itself be very hazardous. This can result in the significant release of dangerous pollutants. The Buffalo Creek flood in 1972 in West Virginia, and the Deepwater Horizon disaster of 2010, off the coast of Louisiana, are but two of many such occurrences. Transporting fuel to locations of use also results in significant toxic emissions. The incidents with the Exxon Valdez (in 1988 near Alaska coast) and the Prestige (2002, off of Spanish NW coast) are two notable examples of oil-laden ships that crashed at sea and spilled millions of gallons of crude oil, causing significant destruction to natural habitats. The economic costs of such disasters are huge, in the multiple billions of dollars. However, it is not just the few major episodes that destroy land and habitat, it is the day to day operation of oil extraction which, often overlooked, sizably impacts large expanses of land as well (Allred *et al.*, 2015).

In energy systems, modern forms of engagement result in potentially disastrous large-scale consequences. The nuclear power industry serves as a great example of how bad things can get. The secrecy and centralization inherent to nuclear energy production coupled with the tremendous amount of radioactive release associated with a major breakdown can potentially create very serious consequences to human health. While the tragedy of Chernobyl (1986) is widely known, the negative effects of the long-term release of radiation from the Hanford Site in Washington state are less so. The Fukushima disaster of March 2011 still has significant impacts today, though much of the information about it is tightly controlled. There are also inherent risks of increasing global nuclear waste; the wastes can be collected and enriched and turned into inherently dangerous nuclear weapons. The possibility of nuclear war continues, unthinkable as it is, as long as the major nuclear powers refuse to abolish nuclear weapons thereby creating a nuclear-free world. The extraction of oil on the continental shelf also comes with great risk — for which the Deepwater Horizon catastrophe is but one such example. Let us not forgot about Bhopal (1984, India) where huge amounts of deadly methyl isocyanate gas were released, a chemical used in the production of carbamate pesticides — implicating the industrialization of India's agricultural sector as the contributor to this loss of ~16,000 lives, not to mention the pain and suffering of thousands of others.

Yet, despite all of these costs, which are not calculated in the cost of fossil fuels at the 'pump' (they are considered 'externalities', to be discussed later), these negative impacts are likely small potatoes compared to the ultimate impacts of catastrophic climate change (C3). The fact that an energy and food system has been created that produces so much CO_2 from burning coal, oil and natural gas, and CH_4 (natural gas "leaks" and animal enteric fermentation) puts all life forms at risk to the vicissitudes of ongoing climate change. Agricultural practices are responsible for approximately one-third of all greenhouse emissions (Nierenberg, 2013). Though this has been documented for years now (the Intergovernmental Panel on Climate Change's (IPCC's) first assessment report was issued in 1990), the U.S. still puts nearly all of its farm dollars into crops that are almost entirely grown (processed, transported, etc.) with fossil fuel inputs. And fossil fuel subsidies have dwarfed the pittances provided to renewable energy over the past 100 years by more than 13 times (Shahan, 2012; more on this in Chapter 6). Annually, government subsidies to the fossil fuel industry amounts to between US$300–500 billion dollars

(Princen *et al.*, 2013). Clearly, a system that would continue to practice such destructive methods is broken and immoral.

In terms of agriculture, turning most of our landscape into GMO products has the potential for catastrophic failure. When the genes of an entire landscape 1,000+ acres in area are exactly the same, the stability of the system will suffer greatly from the introduction of just one pathogen genetically matched to cause the monoculture's destruction. We see some evidence of what catastrophic losses might look like with monoculture agriculture with soy rust and corn grey leaf spot. GMO transgene flow may contaminate non-GMO native species (see evidence in Biello, 2010) undermining the ability of alternative sustainable forms of agriculture to thrive (Garcia and Altieri, 2005).

3.4. Money speaks volumes

Perhaps one of the easiest ways to quantify how big an influence food and energy are on our lives is to look at how much money we spend on them. And while money is about as easily quantified as any other variable, determining precisely how much is spent in these areas remains a bit elusive due to the factor of redundant counting (e.g., energy is purchased to grow, transport, and refrigerate food), as well as ambiguous and elusive externalities. However, a ballpark estimate suffices to establish just how sizeable are their contributions.

On the food front, we spend more than most imagine. In the U.S., consumers, businesses, and government entities spent US$1.46 trillion on food and beverages in 2014 (USDA, 2017b). This equates to US$4,580 a year (or US$88 a week) per person. Interesting, more money is spent on 'away from home' food rather than 'at home' food, reflecting the 'eating out' mentality of modern lifestyles in 'developed' countries (USDA, 2017b). But, as large as the U.S. expenditures are, percentage wise, Americans are spending a smaller fraction of their total income on food, particularly in comparison to those in the developing world. While in the U.S., people spend 7–12% of their income on food, much of the developing world spends in excess of 40% (Washington State University, 2008; Thompson, 2011). Notably, Chinese spend 20% of their money on food, Brazilians, 17%, Indians, 24%, Russians, 34%, and Egyptians a whopping 40% (Thompson, 2011).

Those in the more affluent nations spend a similar amount for energy as they do food. In the US, people (on average) spend ~US$3,052 (in 2012) annually in the residential and transportation energy sectors (USA DOE, 2014). This amount goes up if some of the US$15.7 billion spent in the industrial and commercial sector is added in — clearly some of these expenses are passed on to the consumer. This total also doesn't include what is spent on heating/cooling and lighting public buildings such as city halls and libraries — expenses that are passed on to citizens via taxes (USA DOE, 2014). Of note, the current value does represent a precipitous climb since 1998 when the per capita value was US$1,630 (47% less) (USA DOE, 2014). Notably, the poor in the United States spend a disproportionate amount of their income on energy, in some cases over 50%, particularly, those living in northern climates, though the poor in Alabama and Virginia apparently pay significantly higher proportions as well (Chandler, 2016).

Globally, in a world where 11% live on less than US$1.90 a day (World Bank, 2013) and many millions more live on not much more than that, the bulk of energy expenditures are made by those in the more developed countries and those well-to-do in poorer countries. As of 2011, about US$6 trillion dollars a year were expended globally on energy (Enerdata, 2011), which works out to about US$900 a person annually. This number has more than doubled since 1990 (an estimated increase of 137% in 21 years) (Enerdata, 2011).

None of this is surprising when one considers that petroleum is the largest capitalized industry accounting for 14.2% of all trades of commodities (amounting to US$2.3 trillion annually; Princen *et al.*, 2013). It is also heavily capitalized on a per worker basis with an astounding US$3.2 million invested for each employed person in the petroleum industry (Princen *et al.*, 2013).

3.5. Scale and power

Clearly, energy and food systems are very large entities in our society — the only other expense that rivals these two is medical care. However, it is not just the sheer amount of money involved that is at issue, the actual size of each system is also of great consequence. Over the past 50 years, individual farm sizes have increased to thousands of acres (USDA 2014a) and individual nuclear reactors or coal-fired power plants now serve tens of thousands of households. In this business environment, it has become extremely lucrative to organize/manipulate systems in a way that guarantees future profits. As such, large agribusiness firms have been working feverishly to make food products (as well as all resources used in the preparation of such food) proprietary. Hence, we have seen the development of increasing numbers of seeds that are GMO and patented, along with new pesticides, new synthetic food additives and preservatives; over 90% of corn and soybeans are now grown from GMO seeds. Not only are the seeds patented but so are the pesticides that are used specifically to control pests on this modified crop. And, not surprisingly, the owners of said patents are one in the same. This form of corporate ownership of seeds and chemicals makes the growing of other seeds much more difficult to sustain, particularly in an environment where the crops of patented seeds get the bulk of the federal subsidies. Numerous farmers, especially in areas outside of the U.S., have been unfairly sued because GMO crops have been discovered on land that they are cultivating. The courts have often ruled that this constitutes theft, when actually, it is merely a sign of genetic drift that can be expected to occur naturally when GMO crops have become so ubiquitous on the world's landscapes. In parts of the world, especially India, there have been thousands of farmer suicides (an estimated 200,000 since 1997) in association with corporate threats of land takeover on the basis of 'stolen' seeds, when again, all that has happened is GMO seed has 'moved' into another farmer's land (Shiva, 2013). The chemicals used in association with these GMO crops are potentially incompatible with other species, outside of the one crop (seed) that they were designed to 'protect'. Once released into the environment, these toxic chemicals drift and potentially poison air, water and soil detrimentally affecting other non-target life forms elsewhere. The concentration of ownership of patented GMO seeds (and their associated pesticides) into a few increasingly powerful hands (such as Monsanto, Archer Daniels Midland (ADM), Dupont, Syngenta, and Dow) makes transitioning to alternative forms of agriculture very difficult — as alternatives become less accessible, farmers are more and more compelled to buy into a potentially dangerous form of food production (Joensen *et al.* (2005) describe these issues in the context of Argentina); for more on the problems with GMOs, see section 5.2.

An additional consequence of the scaling up and increased concentration of food ownership is the expansion of grain crops (in acreage and intensity) for the expressed purpose of feeding more animals; currently, some 50% of grain is fed to animals (rather than directly to people) (De Schutter, 2012). Since more profit can be obtained for meat and processed foods than grain itself, the marketing and selling of these products has become the *modus operandi* for multinationals. These changes have also seen a neglect in the growing of nutritious indigenous crops as well (Nierenberg, 2013). Modern diets of residents in 'more developed' countries demand high levels of meat consumption, and as the 'less developed' gain wealth, increased meat consumption is seemingly an expected outcome. According to FAO researchers, annual per capita meat consumption is predicted to increase from 37.4 kg (in 2000) to just under 50 kg in 2050 (Alexandratos and Bruinsma, 2012). As touched on briefly in Chapter 1, this increased meat consumption will have a tremendous impact on the environment. Current meat consumption in the U.S. causes the expanding Dead Zone in the Gulf of Mexico, which in the summer of 2017 covers an area larger than the state of New Jersey (Milman, 2017). These areas become hypoxic (low-oxygen levels) as a result of the excessive nutrient loads they receive from synthetic fertilizers used on industrial farms as well as run-off from the toxic waste generated in concentrated animal feeding operations (CAFOs), which have seen a huge expansion in the past decade, largely under the radar of most city people. These dead areas are anticipated to become larger and more common globally with the expansion of meat consumption, particularly, if methods to produce meat and the feed that 'builds it' do not change. Also, meat and processed food overconsumption has and will continue to expand problems associated with heart disease (primarily from red meat), diabetes and cancer (Nierenberg, 2013).

Sadly, the expansion of CAFOs, and their associated drive to clear more native grasslands to generate soy and corn feedstock, occurs despite a variety of worthwhile efforts on the food sustainability front. Significant efforts have been made by producers to educate consumers of the importance of 'organic', 'free range', 'antibiotic free', 'hormone free', and non-GMO meat (and produce). While many of these claims are not regulated or verified by governmental bodies ('organic' being a notable exception), food labelling of this sort has certainly become more prominent. Yet, despite the appearance of a huge shift to vegan/vegetarian diets in the United States, only 3% of its population describe themselves this way (Milman, 2017).

There is also strong evidence that ever larger and more powerful multinational corporations are working diligently to take ownership of 'natural' seeds companies. Ultimately, these seeds are construed as an economic threat to them, and, given their incredible wealth, they have been moderately successful in gaining a foothold in the natural seeds business. Such an effort continues, and as such, suggest the urgency of prioritizing alternative forms of growing and limiting the ability of corporations to control the future ability of humanity to feed itself. Additionally, nearly all independent organic food companies have been acquired by larger multinational food processors. The infographic compiled by Phil Howard (of Michigan State University) shows how extensive this ownership transfer has been over the past 15+ years (Howard, 2016).

Another negative outcome of the dramatic power of multinational corporations can also be seen in how governments are able to respond to current, emerging and future human and ecosystem duress. Case in point is the recent Colony Collapse Disorder (CCD) of honey bees. While increasing evidence have indicated that a class of pesticides known as neonicotinoids are the primarily source of CCD of bees worldwide (Spector, 2014), the EPA downplays this industrial source, and rather mentions a lot of secondary factors (such as mites and viruses) that aren't created by industry. A second example is the incredible delay, some 21 years, of the national reassessment study of dioxin. The report that was finally released in 2012 had the audacity to say that since exposure to this deadly chemical has been reduced so much over time (ironically, the time corresponding to the delay of the report), human health is now not at risk. Today, we see similar delays and understatements with regard to the research and subsequent findings related to hydraulic fracturing (aka hydrofracking). Despite numerous studies establishing the inherent dangers of toxified drinking water and the release of natural gas into the atmosphere, these practices are largely proceeding unabated — with New York's moratorium being a notable exception. The moratorium is a legal action taken after thousands of protesters forced government agencies to reexamine the available data and make a judgment based on science instead of profits. In most states in the U.S., scientific research is not properly considered when making critically important decisions about human and ecosystem health. A third example is the effort by Big Energy to quash efforts to move towards renewable energies. Farrell (2017) documents how powerful energy companies have attempted, by way of a constitutional amendment, to prevent customer-owned solar energy from being able to compete with them fairly.

3.6. What next?

This record speaks to the inherently unsustainable practices of industrial agriculture that are promulgated aggressively and widely by powerful interests (i.e., the World Bank, IMF, as well as multinational agribusiness). The current system implicitly assumes that non-renewable resources will be available forever and that taxpayer subsidies will continue to allow the dominant flow of purchasing dollars to go to the industrial capitalists who support and continually expand conventional forms of agriculture globally. We find these assumptions poorly grounded in logic and empirical evidence.

Given the grave danger posed by the continued use of fossil fuels and the grand resources required to provide current levels of food and energy, might we just be in a "no-win" situation? Regarding agriculture, fundamentally, 'Can organic, GMO-free agriculture feed the human population, now and in the future?' Many serious studies that have looked at this question have come to the same answer, a resounding 'yes'. While we will provide evidence for this in Chapter 5, a chapter in which we explore the structural elements that can achieve this goal, suffice it to say here that the world is finally coming to realize the urgency of this transformation of agriculture. A report issued by the UN Conference on Trade and Development in 2013, authored by 60 world experts, probably has some of the strongest language, arguing that in response to continued food crises and larger populations in the future *'the world needs a paradigm shift in agricultural development: from a "green revolution" to an "ecological intensification" approach'* [emphasis ours] (UNCTAD, 2013). They add, 'the required transformation is much more profound than simply tweaking the existing industrial agricultural system' and will require a shift to small-scale farming where farmers are directly connected to and have autonomy over their crops, suggesting that major changes are required to meet future needs (UNCTAD, 2013). It is important to recognize that 70% of the foods consumed globally still come from small farmers according to the UN Food and Agriculture Organization (FAO) (Ahmed, 2014). For those that retort, 'it [modern agriculture] produces food more efficiently and in higher quantities than any other food system ever created' or that 'a smaller percentage of the World's people are malnourished or starving today than in earlier human periods', we respond, even if some of these claims have some basis in truth, the means by which these 'accomplishments' were achieved (i.e., the Green Revolution) have come at much too high a price and are flatly unsustainable. Our soils are dying and becoming depleted; about 33% of arable land has been lost since 1975, with erosion rates (on plowed fields) averaging 10–100 times larger than rates of soil formation (Cameron *et al.*, 2015). Our estuaries are too nutrient-rich, driving eutrophication and subsequent loss of existing biodiversity. Our water supplies are full of high levels of toxic chemicals which do extraordinary harm to humans and ecosystems. Through genetic drift, GMO-patented strains are reducing resiliency and genetic diversity on a massive scale and at a rapacious rate, and are actually increasing the use of toxic pesticides (Benbrook, 2012). Farm communities are collapsing socially and economically as farmers are getting a smaller and smaller fraction of the massive expenditures on food production, which get increasingly taken up by transnational agricultural conglomerates who use excess capital to punish those that dare speak against them or their practices, or to sue or buy out seed banks (for documentation watch the two 2008 documentaries, *Food, Inc.* and *The World According to Monsanto*). Furthermore, price fluctuations in food have made tens of millions of economically-challenged people suffer more greatly from abject poverty (Nierenberg, 2013). It is with these realities in mind that alternative paradigms are ripe to be implemented.

Regarding food security, we should do everything in our power to provide sufficient quantities of high quality food to all. While the industrialized and profit-driven food system currently produces more than sufficient calories for the World's human population, it critically fails in four key areas central to this effort. First, it provides too many empty, and dangerous, calories (e.g., high fructose corn syrup derived from GMO corn), causing or exacerbating many health problems (such as heart disease and diabetes). Second, it keeps costs prohibitively high for the poorest individuals (a problem magnified by 'free-trade' agreements, such as NAFTA, and U.S. Farm Bills which subsidize destructive, unsustainable farming practices). Third, it contaminates and simplifies ecosystems, reducing biodiversity. And fourth, its methods result in the loss/waste of one-third — 1.3 billion tonnes annually — of all food produced; shockingly, these food losses amount to nearly US$1 trillion dollars of waste, about US$680 billion (from industrial countries) and some US$310 (from developing countries) (FAO, 2017a). For these reasons, the current dominant paradigm of agriculture is antithetical to our proposed goal for humanity. Clearly, then, the food system (both production and consumption) has to change and must be a central focus of activity for all those committed to creating the OWSP.

However, there are significant barriers to making this an easy transition. Currently in the European Union, 80% of agriculture subsidies and 90% of the agricultural research funding support conventional industrial agriculture (Ahmed, 2014). In the U.S., one finds very few universities offering anything other than preparation for conventional agriculture. Not surprising, many of these institutions are heavily funded by industry and this has a lot to do with setting the course of future research — new faculty are evaluated on the basis of their ability to 'fit' within existing, dominant paradigms. Fred Kirschenmann, currently a Leopold Center Distinguished Fellow, endured a lot of undeserved accusations for his writings and speeches identifying conventional agriculture as the core problem of our society. As such, not surprisingly, many of the alternative approaches, as will be discussed in Chapter 8, are not coming out of academia, but rather out of grassroots or NGOs who are responding to the matter urgently.

Obtaining adequate food and energy is an issue for those who live in areas where food and energy are plentifully produced but improperly distributed, a plausible basis for current conflicts which can be considered 'resource wars'. Several scholars (e.g., Ahmed, 2015 and Klare, 2015) have cogently argued that a great number (if not most) of the conflicts occurring right now have strong drivers in resource shortages, whether actual or created in the interests of profit. And these shortages are largely the result of the increased commodification of and profit obtained from these resources. Sadly, as we will further document in Chapters 4 and 6, the nearly two trillion dollars spent each year for military expenditures (largely to protect/secure these resources, in particular fossil fuel and strategic metal reserves) creates a huge financial drain shortchanging other critical needs, such as education, healthcare, etc. This sad reality has been true for much of the modern era but should not be tolerated, especially not now when sufficient, high-quality alternatives are in abundance, waiting only to be humanely shared, rather than being grossly consumed, hoarded, squandered or wasted because of the current control of food and energy supplies by global elites.

In conclusion, if we were able to tackle the food and energy problems, we would likely be on our way to solving most of the World's current problems. We currently have enough food, we just must begin to share what we have and look at alternative ways to grow it, as current conventional ways are very destructive and unsustainable. Sharing will require us to treat each other as 'brothers and sisters' rather than enemies. Our current industrial agricultural system treats many people on our planet as expendable, exposing them to chemicals in the food they consume or produce, and forcing farmers off their land, in the name of efficiency and profits. Increased population and affluence will increase the need for food, especially meat assuming current modes of production continue, but these modes must be ruthlessly critiqued! Energy resources dominated by fossil fuels are not only currently poorly distributed and not equitable in their delivery, they also need to be rapidly replaced by renewable sources sufficient to provide for increasing demand and to ensure that everyone has enough to reach World standard levels in various yardsticks of quality of life, with one most robustly measured being life expectancy.

Chapter 4

A New Energy Future — Solar

4.1. Introduction

Do we dare, as but two of 7.5 billion now on our planet, prescribe how much and what kind of energy humanity needs? The answer to the question must of course be answered from humanity's collective wisdom. Here we will humbly present our vision, from our perspective as environmental scientists, that will hopefully lead to those most affected by climate change and energy poverty, which now impacts most of humanity, to seriously consider the issue, starting with climate and energy justice advocates who will read this book. We will start with the challenges posed by the appalling weakness of the COP21 Paris Agreement, followed by how global energy poverty can be eliminated in the context of a transition to a global wind/solar energy infrastructure that also has the capacity to become an effective approach to preventing catastrophic climate change (C3) while there is still a window of opportunity. Finally, we will examine the political economic obstacles to realizing this prevention program and how seizing this opportunity is simultaneously a path to the 'other world that is still possible' (OWSP). We note that much of this chapter is derived, but revised and updated, from Schwartzman (2016c) with the exceptions of Sections 4.4, 4.5.1. and 4.5.2.

4.2. The climate and energy challenge since COP21

COP21 concluded its meeting on 12 December 2015 in Paris. This was the 21st meeting of the Conference of the Parties to the United Nations

Framework on Climate Change, a process started in 1992 at the Rio Earth Summit. Climate justice activists generally have a very critical analysis of its outcome, which we largely share. However, it should be noted that this COP meeting was the first at which virtually all countries at least submitted their national plans with regard to climate change, subject to periodic review. And in addition, the Paris Agreement is nearly universal, and as such is a symbolic step towards global cooperation and a more peaceful world, despite all its weakness in confronting the ever closer tipping points to catastrophic climate change (C3). The introduction to The Paris Agreement has strong, even inspiring language:

> 'Recognizing that climate change represents an urgent and potentially irreversible threat to human societies and the planet and thus requires the widest possible cooperation by all countries, and their participation in an effective and appropriate international response, with a view to accelerating the reduction of global greenhouse gas emissions; Also recognizing that deep reductions in global emissions will be required in order to achieve the ultimate objective of the Convention and emphasizing the need for urgency in addressing climate change; Acknowledging that climate change is a common concern of humankind, Parties should, when taking action to address climate change, respect, promote and consider their respective obligations on human rights, the right to health, the rights of indigenous peoples, local communities, migrants, children, persons with disabilities and people in vulnerable situations and the right to development, as well as gender equality, empowerment of women and intergenerational equity' (Adoption of the Paris Agreement, 2015, p. 1–2).

Before getting into the provisions of the Paris Agreement, let us recall the identity of the greenhouse gas emissions and their sources. Roughly 60% of greenhouse gas emissions, dominated by carbon dioxide, come from fossil-fuel use, with coal, natural gas (due to its combustion and methane leakage into the atmosphere), and tar sands oil having the highest carbon footprint (IPCC, 2014). Conventional liquid oil has the lowest carbon footprint, about three-fourths that of coal (Smil, 2003).

The other greenhouse gases derived from human activity include nitrous oxide, the breakdown product of nitrate fertilizer, with carbon dioxide and methane also coming from agriculture, particularly from cattle. This contribution to global warming makes a transition to ecologically-based agriculture imperative. The last anthropogenic greenhouse forcing of significance comes from halocarbons with emissions totaling a few percent of the total.

The Paris Agreement includes a goal of keeping global temperature increase 'well below' 2°C, and to pursue efforts to limit it to a 1.5°C warming above pre-industrial temperatures by 2100, but with no penalties for failing to achieve INDCs (the Intended National Determined Contributions) to curb greenhouse gas emissions over a projected time period.

The Paris Agreement involved 176 nations including the biggest greenhouse gas polluters, China, the U.S. and the EU, having made specific commitments (INDCs) to peak their greenhouse gas emissions as soon as possible and eventually curb them. This requires a review of progress towards increasing their INDCs every five years, in a transparent process (COP21, 2015).

4.2.1. *Significance of the 1.5°C target*

'The fact that the accord prominently mentions the 1.5°C target is a huge victory for vulnerable countries', says Saleemul Huq, director of the International Centre for Climate Change and Development in Dhaka, Bangladesh. 'Coming into Paris, we had all of the rich countries and all of the big developing countries not on our side', states Huq, an adviser to a coalition of least-developed nations. Huq feels that progress has been made, 'In the 14 days that we were here, we managed to get all of them on our side' (Tollfeson and Weiss, 2015).

However, according to Climate Interactive (2017), a major monitor of climate change, based on the current INDCs from the major carbon emitting countries, projected warming by 2100 will be 3.3°C (6.0°F), or 1.3°C above the 2°C warming limit above the pre-industrial global temperature. Note that an earlier assessment by the UN gave a somewhat lower projected warming of 2.7°C (4.9°F), still almost 1°C higher than the 2°C limit (Climate Action Tracker, 2015).

Returning to the issue of the warming temperature target, we find in the Introduction to the Agreement itself:

> 'Emphasizing with serious concern the urgent need to address the significant gap between the aggregate effect of Parties' mitigation pledges in terms of global annual emissions of greenhouse gases by 2020 and aggregate emission pathways consistent with holding the increase in the global average temperature to well below 2°C above pre- industrial levels and pursuing efforts to limit the temperature increase to 1.5°C' (Adoption of the Paris Agreement, 2015, p. 2).
>
> Further, it has been pointed out that, 'According to the IPCC, holding warming to 2°C will probably require emissions to be cut by 40–70% by 2050 compared with 2010 levels, Achieving the 1.5°C target would require substantially larger emissions cuts — of the order of 70–95% by 2050...The new agreement doesn't [fully] take effect until 2020, the chance to achieve the 1.5-degree goal will have already gone, unless all of the world's largest economies dramatically change course' (Reyes, 2015).

There is still a window of opportunity to keep warming below 1.5°C, but it is vanishing fast (Rogeli *et al.*, 2013). A more recent study found that 'anthropogenic emissions need to peak within the next ten years, to maintain realistic pathways to meeting the COP21 emissions and warming targets' (Walsh *et al.*, 2017).

Several leading climate scientists think that the 2°C limit is too high. For example, retired NASA climate scientist Jim Hansen said, 'aiming for the 2°C pathway would be foolhardy', because of projected impacts such as sea level rise and acidification of the ocean (Hansen *et al.*, 2013). His assessment is reinforced by a newly published study in the *Proceedings of the National Academy of Sciences* (Drijfhout *et al.*, 2015). The evidence strongly reinforces the long-term demand of many poor countries for a 1.5°C limit, recognizing that the severe weaknesses in the Paris Agreement make this goal a huge challenge. In an even more recent assessment Hansen *et al.* (2017) argues that:

> 'These considerations raise the question of whether 2°C, or even 1.5°C, is an appropriate target to protect the well-being of young people and future

> generations. Indeed, Hansen *et al.* (2008) concluded that "if humanity wishes to preserve a planet similar to that on which civilization developed and to which life on Earth is adapted, . . . CO_2 will need to be reduced...to at most 350 ppm, but likely less than that", and further "if the present overshoot of the target CO_2 is not brief, there is a possibility of seeding irreversible catastrophic effects." A danger of 1.5 or 2°C targets is that they are far above the Holocene temperature range. If such temperature levels are allowed to long exist they will spur "slow" amplifying feedbacks..., which have potential to run out of humanity's control. The most threatening slow feedback likely is ice sheet melt and consequent significant sea level rise, as occurred in the Eemian, but there are other risks in pushing the climate system far out of its Holocene range. Methane release from thawing permafrost and methane hydrates is another potential feedback, for example, but the magnitude and timescale of this is unclear' (p. 582).

Further, in his commentary on this paper, Hansen (2017) says:

> 'New conclusions we now stress, which were only implicit in the Discussion version, include: (a) Even the aspirational goal of the Paris Agreement, to keep global warming below 1.5°C, is not adequate. (b) A current narrative, that humanity has turned the corner and is moving toward solving the global warming problem, is wrong. Atmospheric greenhouse gases are not only continuing to increase rapidly, their growth rate has actually accelerated rapidly in the past several years... An appropriate goal is to return global temperature to the Holocene range within a century. Such a goal was still achievable in 2013 if rapid emission reductions had begun at that time and if there were a global program for reforestation and improved agricultural and forestry practices. Now climate restoration this century would also require substantial technological extraction of CO_2 from the air. If rapid emission reductions do not begin soon, the burden placed on young people to extract CO_2 emitted by prior generations may become implausibly difficult and costly.'

Some in the media promote even more pessimistic views with respect to the chances of avoiding dangerous climate change (e.g., Mooney,

2017a, citing Rafferty *et al.*, 2017; Mauritsen and Pincus, 2017). However, many assumptions underlie these projections, such as trends in GDP and population, with a subtext ignoring the radical changes necessary in both the physical and political economies (i.e., ecosocialist transition from fossil fuel capitalism). Nevertheless, a growing consensus among leading climate scientists points to a quickly vanishing window of opportunity to avoid C3 (Hansen *et al.*, 2017; Figueres *et al.*, 2017).

China is the World's leading carbon emitter, at emission levels almost double that of the U.S. (which stands in second place). China has committed to leveling off its emissions by 2030 (using carbon emission trading), while the U.S. promised to reduce its greenhouse emissions by 26–28% by 2025 relative to 2005 emissions (Parlapiano, 2015), although of course this commitment is now withdrawn with the current U.S. government's rejection of the Paris Agreement. As Naomi Klein has recently pointed out, citing the assessment of the Tyndall Centre on Climate Research, the U.S. goal falls far short of what is required for even the 2°C goal, which would require emissions reductions of at least 8–10% *per year* (Klein, 2014).

Projected warming in combination with lackluster efforts to cut emissions has created an imminent crisis. This is the reality check for serious activists. Any remaining possibility of preventing warming from reaching a dangerous level will require rapid and radical cuts in global carbon emissions — starting with the highest carbon footprint fossil fuels — and the simultaneous creation of a viable global wind/solar power infrastructure.

The Agreement also includes a commitment of US$100 billion a year in climate finance for developing countries no later than 2025, and for further finance in the future. It states, 'Prior to 2025 the Conference of the Parties … shall set a new collective quantified goal from a floor of USD 100 billion per year, taking into account the needs and priorities of developing countries' (COP 21, 2015, p. 8).

Here are further assessments of the Agreement from two leading climate scientists/activists:

> First: 'It's a fraud really, a fake, … It's just bullshit for them to say: "We'll have a 2°C warming target and then try to do a little better every five years." It's just worthless words. There is no action, just promises. As long as fossil fuels appear to be the cheapest fuels out there, they will

continued to be burned' (Jim Hansen, retired NASA climate scientist, in Milman, 2015).

Second: 'Since 2009, U.S. State Department chief negotiator Todd Stern successfully drove the negotiations away from four essential principles: ensuring emissions-cut commitments would be sufficient to halt runaway climate change; making the cuts legally binding with accountability mechanisms; distributing the burden of cuts fairly based on responsibility for causing the crisis; and making financial transfers to repair weather-related loss and damage following directly from that historic liability. Washington elites always prefer "market mechanisms" like carbon trading instead of paying their climate debt even though the U.S. national carbon market fatally crashed in 2010' (Patrick Bond, leading climate justice activist from South Africa, in Bond, 2015).

Further focusing on his own country, with relevance to the later discussion on the BRICS, Bond adds:

'"South Africa played a key role negotiating on behalf of the developing countries of the world," according to Pretoria's environment minister Edna Molewa, who proclaimed from Paris "an ambitious, fair and effective legally-binding outcome". Arrogant fibbery. The collective Intended Nationally Determined Contributions (INDCs) — i.e. *voluntary* cuts — will put the temperature rise at above 3 degrees. From coal-based South Africa, the word ambitious loses meaning given Molewa's weak INDCs–ranked by Climate Action Tracker as amongst the world's most "inadequate" — and given that South Africa hosts the world's two largest coal-fired power stations now under construction, with no objection by Molewa. She regularly approves increased (highly-subsidized) coal burning and exports, vast fracking, offshore-oil drilling, exemptions from pollution regulation, emissions-intensive corporate farming and fast-worsening suburban sprawl' (Bond, 2015).

A powerful critique of the Paris Agreement comes from indigenous communities as emphasized by Holleman (2017). She argues the following:

'All this points to imperialism as central to the ecological crisis…It shows that attempting to address ecological crises without challenging the social

> order serves to reinforce the undemocratic social relations of capital, simply displacing or shifting crises socially, geographically or temporarily. Understanding why this is the case under capitalism should be an urgent goal of all environmentalists. If we don't understand the historically specific political and economic dynamics of the society in which we live, we cannot effectively organize, even for reform. Committed organizers for social change, past and present, have understood this. Another urgent and related goal should be understanding why the mainstream environmental movement, in spite of claims to support "environmental justice", has so far failed to fight consistently in solidarity with the most oppressed communities for a more radically democratic society — for an alternative to a system with such toxic priorities. This lack of solidarity has had grave consequences for the efficacy of the movement and hence for the planet. Along the same lines, why have prominent environmentalists helped political elites declare the Paris agreement a victory and obscure (1) the lack of substantive ecological content in the agreement, and (2) its oppressive and imperialistic implications?' (p. 164).[4.1]

Nevertheless, the ongoing implementation of the Paris Agreement should be seen as a critical venue for struggle to prevent climate catastrophe (e.g., see Dobson, 2018). It is heartening that climate scientists are now

---

[4.1] Holleman (2017) quotes the following as said by Calfin Lafkenche, Mapuche leader from Chile, at the 15th session of the UN Permanent Forum on Indigenous Issues in May 2016: 'We are here today in the UN to stop the offensive of the Green Economy and its market systems of carbon trading, carbon offsets, the Clean Development Mechanism, and REDD+, which constitute a new form of colonialism and have caused conflicts, forced relocation, threats to the cultural survival and violations of the rights of Indigenous peoples, especially the rights to life, to lands and territories, and to free, prior and informed consent' (p. 162)

'The Paris Agreement is a trade agreement, nothing more. It promises to privatize, commodify and sell forest and agriculture lands as carbon offsets in fraudulent schemes such as REDD+. These offset scams provide financial laundering mechanisms for developed countries to launder their carbon pollution in the Global South. Case-in-point, the United States' climate change plan includes 250 million megatons to be absorbed by oceans and forest offset markets. Essentially, those responsible for the climate crisis not only get to buy their way out of compliance but they also get to profit from it as well' (Indigenous Environmental Network, 2016).

focused on modeling scenarios of limiting warming to below the 1.5°C goal (e.g., Xua and Ramanathan, 2017; Millar *et al.*, 2017; Goodwin *et al.*, 2018).

4.2.2. *New evidence for climate change impacts*

It is claimed that new alarming evidence of accelerated warming potential has emerged from the threat of rapid melting of Arctic permafrost triggering the sudden release of methane to atmosphere, citing research pointing to a potential 50 billion ton burst of methane into the atmosphere from the decomposition of methane hydrates from thawing Siberian Arctic permafrost (Jamail, 2017). Permafrost loss and consequent methane release into the atmosphere are rising faster than previously modeled (Chadburn *et al.*, 2017). This would be equivalent to a sudden global warming from an increase of 81 ppm in atmospheric carbon dioxide level (it is now ~400 ppm). Further evidence of the warming-induced release of carbon dioxide from Arctic permafrost comes from studies in the Alaskan tundra (Commane *et al.*, 2017). However, other recent research makes the case that an imminent runaway methane-driven greenhouse is unlikely (Bock *et al.*, 2017; Hong *et al.*, 2017).

New research also points to an accelerating trend to severe weather events as the climate warms (Liu, 2017) as well as the direct impact of heat stress on people not in air-conditioned environments:

> 'An increasing threat to human life from excess heat now seems almost inevitable, but will be greatly aggravated if greenhouse gases are not considerably reduced' (Mora *et al.*, 2017, abstract).

A recent study (Matthews *et al.*, 2017) concludes,

> 'With only 1.5°C of global warming, twice as many megacities (such as Lagos, Nigeria, and Shanghai, China) could become heat stressed, exposing more than 350 million more people to deadly heat by 2050 under a midrange population growth scenario. The results underscore that, even if the Paris targets are realized, there could still be a significant adaptation imperative for vulnerable urban populations.'

The growing threat to global biodiversity has also been highlighted (Jamail, 2017b). New evidence has emerged for an accelerating rate of vertebrate extinctions, with a recent study concluding:

> 'Earth is experiencing a huge episode of population declines and extirpations, which will have negative cascading consequences on ecosystem functioning and services vital to sustaining civilization. We describe this as a "biological annihilation" to highlight the current magnitude of Earth's ongoing sixth major extinction event' (Ceballos *et al.*, 2017, abstract).

Moreover, there is now alarming new evidence for declining oxygen in the global ocean and coastal waters as a result of global warming combined with nutrient inputs, with potentially major changes in ocean productivity, biodiversity, and biogeochemical cycles (Breitburg, *et al.*, 2018). New research points to the previously underestimated threat of sea level rise (Mooney, 2017b) and intense thunderstorms in the tropics and subtropics (Singh *et al.*, 2017). Parenti (2017) provides an up-to-date summary of what is in store for humanity as soon as the next decade if C3 prevention is not implemented in time.

In spite of these very serious warning signals on one hand and the often cited evidence for exponential growth in the renewable energy sector on the positive side (e.g., Blakers, 2017), we invite the reader to look hard at all the evidence. A turning point in implementing a prevention program for C3 is still within reach, with likely less than a decade left to act, but we are not there yet. Sweeney and Treat's (2017) thorough review is sobering while also offering hope for the implementation of a prevention program for C3 in the ever diminishing time we have left.

4.2.3. *A transition is not only possible, but imperative*

Given the global North's historic responsibility for the threat of C3, the transfer of wind and solar capacity to the global South from the global North is imperative. With 1–2% of current annual consumption of energy (85% derived from fossil fuels) being used for wind and solar power creation per year, a global-scale transition can be achieved in no more than thirty years, with the complete elimination of anthropogenic carbon emissions (derived from energy consumption) into the atmosphere, and the provision of the minimum per capita energy consumption level required for state-of-the-science life expectancy level for all. Further, if this transition commences in the near future it would maximize the probability of achieving a less than 2°C increase over the pre-industrial level, with a potentially less than 1.5°C limit by 2100 (see e.g., Schwartzman *et al.*, 2016).

Thus, a rapid phaseout of fossil fuels, starting with those with the highest carbon footprint, coupled with a full transition to a global wind/solar power infrastructure, should be a global objective. This approach even has the potential to keep overall warming below 1.5°C, in a roughly 25-year transition, if it begins robustly in the near future and is combined with carbon sequestration from the atmosphere.

4.2.4. *What is the way forward for climate justice?*

Tokar (2014) has provided an illuminating history of the climate justice movement. Since this assessment, the challenges to this movement have greatly increased. But rather than immobilizing the climate justice movement from the recognition of the huge challenges unaddressed in the COP21 Agreement, indications so far point to a reenergizing process as a result. This movement is taking inspiration from the rejection of the XL Keystone pipeline by President Obama (but with the struggle continuing since the reversal of this decision by the Trump administration), and is continuing opposition to energy projects with high environmental/ecological footprints. We should also recognize the critical actions of cities around the world to take more aggressive steps to curb their greenhouse gas emissions and transition to renewable energy supplies. But the problem is larger than COP21 and the climate denialism of the Trump administration. Unless the climate justice movement succeeds in broadening its scope by first identifying the biggest obstacles to implementing a prevention program in time to avoid catastrophic climate change, humanity will face a very bleak future with only partial adaptation available as an option.

4.2.5. *Energy poverty and its elimination*

Primary energy consumption, now corresponding to a global power value of 18 trillion watts (TW), is the total energy produced, including waste heat, by society. Watt is a unit of power which equals energy supply divided by the time over which it is received/produced. Hence the global primary energy consumption in one year is now 18 TW-years (or 158 PWh, where P = peta (10^{15}) and h = hours, the unit typically used). The consumption per person in a nation is the total energy consumed divided by its population and captures how much energy is being used for health, education and a multitude of other human activities. We submit that at least before state-of-the-science energy efficiencies are globally implemented, reaching a minimum 3.5 kW/person in primary energy consumption is necessary but not sufficient for acquiring the highest life expectancy (Smil, 2008, p. 346), noting that several petroleum-exporting countries in the Mid-East as well as Russia fall well below this value (see Table 4.1 and Figure 4.1). Life expectancy for the United States is likewise below most industrial countries of the global North; now tied with Cuba, the U.S. ranks 34th in the World. Income

Table 4.1. Energy poverty in Africa, primary power consumption

	Life Expectancy (at birth, yrs)	**Power per capita (kW/person)**	**Gini Index (Year of data)**
Libya	73	3.8	0.35 (pre-2012)
South Africa	63	3.6	0.63 (2013 est.)
Nigeria	55	1.0	0.43 (2005)
Mozambique	58	0.6	0.46 (2008)
For comparison:			
Cuba	79	1.4	0.30 (pre-2012)
United States	79	9.2	0.41 (2010)
Iceland	83	23.2	0.28 (2006)

Sources: Life Expectancy: WHO (2016), 2015 data.
Power per capita: World Bank (2014), 2014 data.
Gini Index: Wikipedia (2016b); year of most recent data available; except for Libya and Cuba: posted 2012.

inequality is robustly correlated with bad health and must be reduced to achieve the world standard life expectancy and quality of life (Kawachi and Kennedy, 2006; Wilkinson and Pickett, 2009). The standard measure of income inequality is the Gini coefficient (named after its designer), which ranges from 0 (lowest level of inequality) to 1 (highest level).

Energy poverty in Africa is widespread, for example, with Nigeria and Mozambique consuming 1.0 and 0.6 kW/person respectively, and having corresponding life expectancies of 55 and 58 years (see Table 4.1). Two exceptions are Libya (pre-2012) and South Africa with 3.8 and 3.1 kW/person, respectively, and corresponding life expectancies of 73 and 59 years. Significantly, South Africa has very high, and Libya very low, income inequality, as measured by their Gini index.

Supplying the minimum 3.5 kW/person for the present World population of 7.35 billion people would require a delivery equivalent to 26 TW noting again the present delivery equal to 18 TW, 31% lower.

Ted Trainer (2014), a prominent advocate for a global reduction in energy consumption, is skeptical of the 3.5 kW/person minimum, arguing that a 'satisfactory quality of life does not require 3.5 kW [per capita]'. But what does 'satisfactory' mean? A lower life expectancy than the highest now achievable, with the implication that most of humanity must settle for less while the privileged elites in the global North get the best health care? In contrast, we offer the following imperative: every child born on our planet has the right to the state-of-the-science life expectancy now shared by a few countries in the global North, not simply a 'satisfactory' quality of life (Schwartzman, 2014b, 2014c).

The 3.5 kW/person minimum is an approximate but plausible value for a first-order global goal for a solar transition commencing in the near future. Life expectancy is a more robust measure of quality of life than the Human Development Index (HDI), which includes gross domestic product *per capita* in its calculation. Note that Cuba, a country that does remarkably well in health and education has a life expectancy now equal to the United States, while Cuba's energy consumption *per capita* is 1.4 kW/person, and the U.S.'s is 9.2 kW/person. Cuba, a country on the ecosocialist path with respect to agroecologies and renewable energy development in rural areas (Murphy and Morgan, 2013), suffers from energy poverty and its associated scarcities, with the U.S. embargo having a decisive role. Judging from the

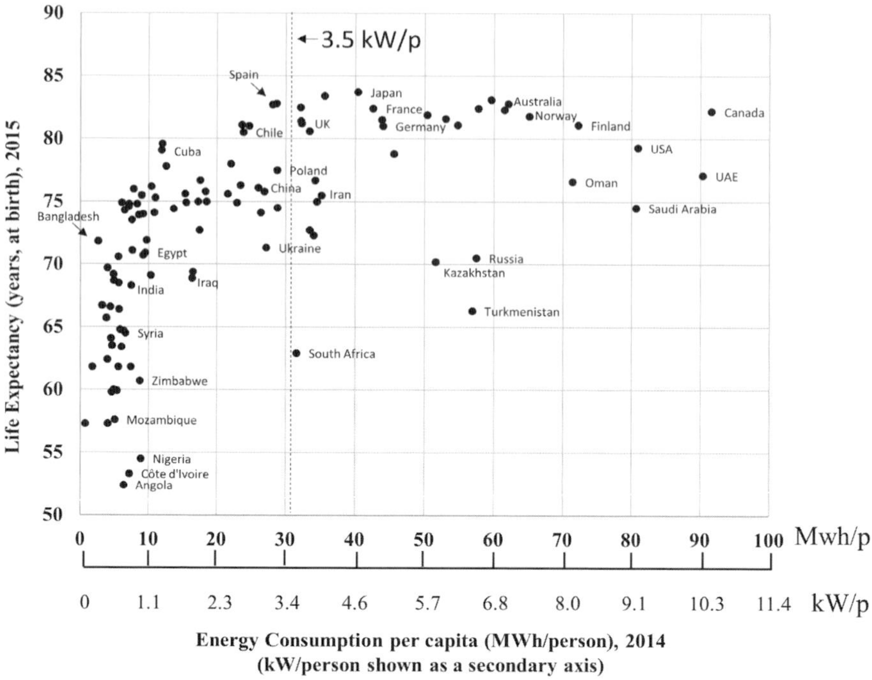

Figure 4.1. Life expectancy versus energy consumption (per capita)

Sources: Life expectancy (WHO, 2016); Energy consumption (World Bank, 2014).

graph of life expectancy versus energy consumption per capita by nation (Fig. 4.1), the minimum necessary for the present World's highest life expectancy is now plausibly between 2.8 and 3.5 kW/person, but closer to the higher value for a global average, with the highly urbanized Hong Kong near the lower value (D. Schwartzman and P. Schwartzman, 2013).

A value closer to 3.5 kW/person for a global average better captures the impact of regional climate on energy consumption, both for air conditioning in very warm and heating in very cold climates. Note, for example, that Iceland (Arctic climate) ranks 9th in the World with respect to life expectancy, yet consumes 22 kW/person. As the global climate warms, the energy demand for the global South, where most of humanity lives, is expected to increase, out-weighing a decrease from lower heating needs in high latitudes of the global North.

While the U.S. and several other countries, with wasteful excess *per capita* consumption, surely need to reduce their energy consumption, and recognizing that pockets of energy poverty exist even in the global North, most of the global South requires a significant increase to achieve 'state of the art/science' quality of life.

A shift to wind and solar-generated electricity as an energy source should ultimately reduce the required power level by roughly 30% once a global system is created, given the greater 2nd-law efficiency of solar versus fossil fuels, i.e., to do the same work solar requires less energy (Jacobson and Delucchi, 2009; Jacobson *et al.*, 2014). For example, an electric car charged by solar-produced electricity is more efficient (in transporting the rider) than fossil fuel combustion in a gasoline-powered vehicle with its large fraction of generated energy going to heat the engine. As a result of this Second Law efficiency, the impact of solar generated electricity as well as plausible increases in the work to energy ratio from other technological efficiencies, the required global energy level for optimal human development should ultimately decrease upon the completion of a global solar transition (see later discussion in this chapter).

4.2.6. *Why wind/solar energy should be used to eliminate energy poverty*

Fossil fuel combustion and the lack of clean energy are killing millions of people across the globe every year, especially in Asia. As a result, recent estimates point to 3 to 7 million per year dying from air pollution (UNEP, 2014; Lelieveld *et al.*, 2015), with 1.6 million in China alone (Rodhe and Muller, 2015).

In the U.S., numerous epidemiological studies have found a strong link between air pollution and childhood asthma, cancer, and cardiovascular disease. Indeed, even climate change deniers can appreciate the shortening of life from air pollution, though one should not underestimate the depth of delusion driven by the right-wing corporate sector regarding both climate change and environmental health. The fierce attack on the EPA's ozone standard is a prime example. Regardless, connecting the dots between fossil fuel consumption and adverse health impacts from air pollution is a fertile strategy for highlighting environmental injustice and strengthening the movement against climate change both in the U.S. and especially in the global South.

4.2.7. *Wind/solar transition as a C3 prevention program*

Mark Jacobson's group at Stanford University has demonstrated the technical feasibility of a rapid global transition to renewable energy (Jacobson and Delucchi, 2011; Delucchi and Jacobson, 2011; Jacobson *et al.*, 2017b; Jacobson, 2017a; 2017b).[4.2]

We modeled the global solar transition with simulations that included values for the energy return over energy invested (EROEI) for state-of-the-science wind/solar technologies, i.e., the amount of energy technologies such as a photovoltaic array or wind turbine generate in their usable lifetimes divided by the energy needed to construct and maintain them (P. Schwartzman and D. Schwartzman, 2011). State-of-the-science EROEI

[4.2] The Jacobson group modeling of a 100% transition to wind, solar and hydropower for the U.S. was challenged by Clack *et al.* (2017), with a simultaneous rebuttal by Jacobson *et al.* (2017a), followed up by Jacobson (2017a). One point of contention was whether the addition of hydropower turbines could serve as a backup energy supply. In Jacobson *et al.* (2017a), as well as in their latest version of their modeling of a transition to energy supplies derived from Wind, Water, and Sunlight (WWS) (Jacobson *et al.*, 2017b), a case is made for an alternative backup utilizing concentrated solar power (CSP) and batteries. In the latter modeling study, they project for 2050 a global end use of 11.84 TW; using the assumed 23% improvement in efficiency from electrification, and 6.98% end-use efficiency beyond a B.A.U. scenario. Is a global level of 11.84 TW sufficient to wipe out energy poverty (for their projected 8.80 billion people, this corresponds to 1.35 kW/person)? Projected 2050 values for China and India are 1.32 and 0.63 kW/person respectively, while the U.S. would have 3.05 kW/person. We conclude that energy poverty is the outcome in this modeling for China and India. This study assumes a baseline end-use load of 12.1 TW for 2012, for the same year the primary global energy consumption is 18.36 TW (IEO, 2016), the same global level assumed in the previous discussion of life expectancy versus energy consumption per person. Thus, as a first approximation, the computed kW/person levels can be evaluated with respect to our present values. However, as Jacobson *et al.* (2017b) points out, with respect to the impact on life expectancy, the very significant reduction in premature deaths from air pollution in the WWS transition should be taken into account. An additional challenge not transparently considered in this modeling is the incremental energy needed for climate mitigation/adaptation which would significantly increase the required global energy demand.

values of current technologies for wind turbines range from 20 to 75; for photovoltaics, greater than 10; and for concentrated solar power (CSP), 7 to 40. Some still argue that the EROEI ratios of these wind and solar technologies are significantly lower than these values, but these claims have been systematically rebutted (see discussion in Schwartzman, 2014); even Charles Hall, influential in arguing along these lines, e.g., Hall et al. (2014) cites a ratio of 18 for wind from 2010 publications (Murphy and Hall, 2010; Kubiszewski *et al.*, 2010). Particularly illuminating is the critique from the Bountiful Energy Blog (2017) (see Appendix 2). We can anticipate near future revolutionary developments in these technologies, with significantly higher EROEI ratios, thereby making possible a faster transition toward using even less fossil fuel than building a global energy infrastructure using present technologies. A new theoretical study finds significant potential for harvesting outgoing thermal radiation for renewable energy applications, achieving both high efficiency and high operating power density simultaneously (Buddhiraju *et al.*, 2018). Already the first floating wind farm is being constructed off the coast of Scotland (Harrabin, 2017b), a precedent for more efficient collection of wind power in the sea.

Our report (P. Schwartzman and D. Schwartzman, 2011) was apparently the first study which computed the necessary non-renewable energy (mainly fossil fuel) needed to create the renewable capacity in a solar transition scenario. The critical factor that leads to exponential growth of this renewable energy supply is the feedback of energy from the growing renewable capacity back into the physical economy to create more of itself.

As previously noted, in our modeling study, assuming a composite EROEI of 20–25, we demonstrate that with only a small fraction of the current annual consumption of energy being used for wind/solar power creation per year, we can achieve a global-scale transition in no more than 30 years, with the complete elimination of anthropogenic carbon emissions into the atmosphere derived from energy consumption, and the provision of the minimum *per capita* energy consumption level required for state-of-the-science life expectancy level for all. In a follow-up study, we show that this solar transition has the capacity to reduce the atmospheric carbon dioxide level below 350 ppm in this century (D. Schwartzman and P. Schwartzman, 2013).

The material resources and land area needed for global solarization are already within reach:

(1) If 15% of present World rooftop area were to be used to site *photovoltaics* with an assumed conversion efficiency of 20%, the current global electricity power capacity would be created. (This calculation assumes a conservative solar radiation flux corresponding to the United Kingdom. An estimate of global rooftop area is 3.8×10^{11} m^2 from Akbari *et al.*, 2009). The photovoltaic industry is already taking seriously the challenge of creating terawatt scale infrastructure (Weber *et al.*, 2017).

(2) Research in the last decade concudes that a *global wind turbine infrastructure* could deliver several times the present global energy consumption, without closing off most of the land where it is sited to other uses (e.g., farming), while having small impacts on regional climate (Lu *et al.*, 2009; Marvel *et al.*, 2013; Grassi *et al.*, 2015; Kleidon *et al.*, 2015). However, considering the constraints on extractable wind power on land as a result of decreasing kinetic energy as the number of wind turbines increase, most of this potential will be from wind turbines sited in the open ocean (Possner and Caldeira, 2017). For example, the latter authors found in their modeling that a sufficiently large wind farm in the North Atlantic could supply comparable power to the present level of 18 TW. Consider the following example, suppose 5 MW capacity wind turbines supply all this energy, with a 35% capacity factor, i.e., the delivered power is 35% of the 'nameplate' capacity of 5 million watts (MW) because of variation in actual wind speed. Then producing 25 trillion watts (TW) of energy would require the manufacturing of 15 million wind turbines in 25 years, assuming the lifespan of this technology exceeds this timespan. In comparison, 51,473 MW of wind turbine capacity was installed globally in 2014, equivalent to 10,300 turbines with 5 MW capacities (Global Wind Energy Council, 2017). Nevertheless, the production of 15 million wind turbines (5 MW each) is plausibly within the technical capacity of the global economy, noting that 90 million cars and commercial vehicles were globally produced in 2014 alone (International Organization of Motor Vehicle Manufacturers, 2014)! It should be noted that state-of–the-technology capacity factors now

commonly reach 40–50%, requiring proportionately fewer turbines to supply the same energy (IRENA, 2015, their Fig. 4.12). Of course, a wind/solar transition using the main three technologies listed here would require even fewer turbines. We will shortly revisit the issue of just how much energy humanity will need in the coming decades, taking into account other challenges besides energy poverty.

(3) *Concentrated Solar Power* (CSP) in the Sahara could supply the current global electricity consumption on less than 6% of the Saharan land area, not that CSP should be only sited in the Sahara of course, as proposed in The Trans-Mediterranean Renewable Energy Cooperation (TREC) Project (European Business Council for Sustainable Energy, 2017).

Finally, it is important to note that the generation of electricity from renewable energy is now cheaper than both fossil fuel and nuclear-based power (Hill, 2017) and this cost differential is expected to grow with ever more efficient solar and wind technologies. The latter technologies have long been cheaper than fossil-fuel and nuclear power taking into account externalities never expressed in cost. In addition, a recent study points to much lower lifecycle GHG emissions from wind and solar technologies than fossil fuel (Pehl *et al.*, 2017).

4.2.8. *Too much land for solar/wind?*

A common argument against a solar/wind transition away from fossil fuels/nuclear power is that too much land is needed for this transition. As we have just shown, a combination of offshore wind, CSP in deserts and photovoltaics on roofs can supply the World in this transition. Further, even wind farms on land occupy a small fraction of the area that can be used for agriculture. Life-cycle energy density analysis typically gloss over these considerations (e.g., Cheng and Hammond, 2017), while Jenkins (2015) gives a more balanced assessment. Considering the land already dedicated to fossil-fuels, Allred *et al.* (2015) estimate that 3 million hectares of land ecosystem was lost from oil/gas development in Central North America (U.S. and Canada) since 2000. Commenting on this paper, according to Mahan (2015) says:

> 'And that's just oil and gas — think about the upstream land destruction for coal-fired power plants (mountain top removal for coal mining comes to mind) and nuclear reactors. We've made some calculations showing the natural resources saved by using wind power as opposed to coal, gas or hydropower in a previous post. The study in *SCIENCE* did not include land losses associated with the nearly 300,000 miles of natural gas pipeline across the Lower 48 States, nor did it calculate land loss associated with the corresponding fossil fuel power plants nor fossil fuel refineries. The study also didn't look at oil and gas land use east of the Great Plains (the Marcellus shale wells in Pennsylvania were excluded, so was all of Louisiana), nor west of the Rocky Mountains (California and Alaska were also excluded) so it's fair to say the land loss estimates are very conservative.'

Note that 3 million hectares is 0.4% of U.S. land area, excluding California and Alaska. This estimate is a good minimum because while Central Canada is included in the Allred *et al.* (2015) study, its oil production was a service to the U.S. since over 3 million barrels of oil/day produced mainly in Central Canada were exported in 2016, primarily to the U.S. (Mawji, 2016), with total production varying month by month between 2.8 and 4 million barrels/day (Canada Crude Oil Production. 2017).

How does this land area impacted by oil/gas production compare to the land needed for a 100 percent conversion to wind/solar power Jacobson *et al.* (2015) conclude from their U.S. modeling:

> 'The new footprint over land required will be ~0.42% of U.S. land. The spacing area between wind turbines, which can be used for multiple purposes, will be ~1.6% of U.S. land...Table 2 [in this quoted study] indicates that the total new land footprint required for the plans, averaged over the U.S. is ~0.42% of U.S. land area, mostly for solar PV power plants (rooftop solar does not take up new land). This does not account for the decrease in footprint from eliminating the current energy infrastructure, which includes the footprint for mining, transporting, and refining fossil fuels and uranium and for growing, transporting, and refining biofuels" (p. 2097).

In their 139-country study (Jacobson *et al.*, 2017b), with a 2050 target date, the land footprint plus spacing area required (in addition to existing installations as of 2015) is 1.2% of these countries' land area, with most devoted to onshore wind power. Thus, we can conclude that a wind/solar transition would actually impact less land than the present legacy and infrastructure of fossil fuels/nuclear power.

4.2.9. *What fossil-fuel is optimal for the solar transition?*

Considering only the greenhouse gas (GHG) footprint of carbon dioxide derived from combustion, coal (anthracite) has twice the impact of natural gas (methane) and 1.4 times the impact of diesel fuel/heating oil (US EIA. 2018). But natural gas is not the "bridge" fuel to renewables, as claimed by some, since the GHG footprint of both shale gas and conventional natural gas is greater than coal because of methane leakage (to the atmosphere), considering a 20-year time period for comparing warming potential of methane to carbon dioxide (Howarth, 2014). Hence, mainly because of its lower GHG footprint compared to coal, the preferred fossil fuel to make a solar transition is petroleum, but only conventional oil, *excluding* the higher GHG footprint tar sands and natural gas, as well as dangerous drilling in deep water continental shelves and other problematic locations such as rainforests. We estimate that if a robust solar transition begins in the near future, it can be completed in 20 to 30 years using less than 20% of the proven conventional reserves of petroleum, with the potential to limit warming to 1.5°C (Schwartzman *et al.*, 2016; P. Schwartzman and D. Schwartzman, 2017). The modeling study of Sgouridis *et al.* (2016) gives similar results. The requirements of petroleum and this timescale will be reduced as higher EROEI wind/solar technologies are developed and put in place in this transition. Since GHG emissions increase with oilfield age (Masnadi and Brandt, 2017) minimizing the use of even conventional oil as an energy source for this transition is imperative. At the culmination of this solar transition, a global increase in energy would be delivered to the world, not a decrease, with many countries in the global North (such as the U.S.) decreasing their wasteful consumption, while most of humanity living in the global South receive a significant increase and reach the rough minimum required for state-of-the-science life expectancy levels.

4.2.10 *Beyond Extractivism?*

Further, with the solarization and decarbonation of global energy supplies, recycling and industrial ecologies powered by wind/solar power should greatly reduce the need for mining. For example, recycling rates of rare earth metals, including neodymium used in wind turbines, is currently very low, less than 1% (Reck and Graedel, 2012). Increasing these rates, as well as implementing alternative technologies, could greatly reduce mining for these and other metals used in modern technologies. In this regard, can we anticipate a transition to a post-extractivist future, parallel to the wind/solar transition? Schwartzman and Saul (2015) propose just this future:

> 'Hollender (2015) has brilliantly outlined the alternatives to real existing extractivism, with the worst being "predatory." Following her terminology, we advocate a transition to an initial "sensible extractivism" followed by an "indispensable extractivism" phase. The characteristics of sensible extractivism include deliberate selecting of extractive projects that meet strong environmental and social criteria, such as those outlined in national and international legislation. For example, extractive projects that risk irreversible environmental damage or species lost should be immediately halted. During this phase, a country would implement additional macroeconomic reforms such as price correction for primary materials, tax and royalty reforms, subsidy reductions, etc. The transition to an indispensable phase of extractivism involves eliminating all extractive activities that cannot be directly linked to human needs and quality of life improvements. Each policy phase would require complementary sociocultural measures to facilitate the long transformation of consumption patterns and "materialistic values." Critical components of this transition include implementing "Buen Vivir" (food security and sovereignty; autonomy in education, governance, and justice; and "making Mother Earth a subject with rights") and the solidarity economy (Hollender 2015, 86–88). In the final phase of ecosocialist transition we envision indispensable extractivism fading away as solar power capacity increases, that is, more solar, more recycling, less mining … Finally,

mining on Earth may cease entirely as mining on the Moon becomes feasible...'

A growing renewable energy capacity should be dedicated to the cleanup and repair of the biosphere after many years of assault by the Military Industrial Complex (MIC), as well as the imperative need to sequester carbon dioxide from the atmosphere to reach the safe limit of less than 350 ppm. In other words, a global solar power infrastructure can increase material production and consumption as needed without the negative impacts now witnessed by unsustainable capital reproduction powered by fossil fuels and nuclear energy.

4.2.11. *Can this wind/solar transition provide energy on demand to global society?*

Baseload is the backup supply of energy when a particular energy technology is not operating at full capacity. Commonly, supporters of continued reliance on fossil fuels and/or nuclear power raise the objection that wind/solar power cannot meet the challenge of baseload. But this claim is misleading. Already available reliable and relatively cheap storage technologies, along with tapping into hydropower and geothermal energy, will facilitate the expansion of renewables and provide baseload power capacity. New advances in battery storage point to the use of common rather than rare elements (e.g., Science News, 2015). A potentially promising approach is to use the obsolete infrastructure from the fossil fuel era, e.g., compressed air storage in shutdown coal-fired plants (Spector, 2017). However, a big enough array of turbines, especially offshore, can likely generate a baseload supply without the need to supplement it with separate storage systems. Further, the progressive expansion of a combined system of wind, photovoltaics, and concentrated solar power in deserts will generate a baseload, simply because the wind is blowing and the sun is shining somewhere in the system if linked to a common grid (Archer and Jacobson, 2007; Kempton *et al.*, 2010). Meanwhile during the transition, baseload would be backed up by petroleum, with the highest carbon footprint fuels phasing out first, on the way to a completely wind/solar global energy infrastructure (D. Schwartzman and P. Schwartzman, 2013).

4.2.12. *Is nuclear energy an alternative for carbon-free energy?*

Expansion of nuclear energy, including a reincarnation of fission-powered reactors with new technology, is not the best option to mitigate global warming, nor will it plausibly avoid the well-known negative environmental and health impacts and risks of this energy source. It has often been argued that nuclear power has no effect on climate change because it does not produce GHGs. But this assertion is not correct since the existing energy infrastructure (mainly fossil fuels) powers all aspects of the nuclear fuel cycle, including the mining and enrichment of uranium, the decommissioning obsolete reactors, and the construction of new reactors (Smith, 2012; Van Leeuwen, 2013). Hence, carbon dioxide emissions do result, even though virtually none occur during the actual production of electricity by the nuclear power plants themselves. The time necessary to create nuclear power replacing existing energy is on the order of a decade, significantly longer than wind/solar plants for equivalent energy supply capacities. Large wind farms can be constructed in about a year. A case in point is the Hinkley Point nuclear power reactor in Somerset, U.K. now under construction, with a projected time to completion on the order of a decade, and a minimum cost of £24.5 billion. Studies have shown that for the same investment at least six times the power generation capacity could be created with wind turbines, and in a much shorter time (Landberg, 2015). An aggressive reduction of greenhouse gas emission with a chance of avoiding C3 requires as rapid a replacement of fossil fuels as possible.

But what about the long-promised delivery of fusion power? We can't wait decades for this energy source to materialize, in contrast to the fusion reactor already being tapped: the Sun at a safe distance of 93 million miles from Earth. The latest forecast for harnessing Earth surface fusion power is beyond 2050 (Cartlidge, 2017).

4.3. Political economic obstacles to C3 prevention

Within the climate justice movement, especially for leaders such as Bill McKibben and Naomi Klein, there is an ongoing lack of serious confrontation of the critical obstacles posed by militarism and imperialism, both of which are integral to real existing capitalism in the 21st Century.

Aside from the 'No War, No Warming' initiative of 2007 during the Iraq war, the climate justice movement has not spoken to the critical role of militarism and imperialism in both contributing to climate change and blocking the implementation of a prevention program.

The major culprit in climate change is not only the fossil fuel industry but its home, the MIC, which sits at the center of capital reproduction. With the increasing awareness of the actual practices of the National Security State, Eisenhower's MIC, now a hundred times bigger, should be reconceived as the 'Military Industrial (Fossil Fuel, Nuclear, State Terror and Surveillance) Complex'. The MIC is not simply a lobby as Eisenhower conceived, but an integrated system of production, powered largely by fossil fuels, even as the actual military goes 'green', e.g., uses solar power on its bases around the World. Rosa Luxemburg was a seminal Marxist activist and scholar in the early 20th Century. Rein (2016) makes critically important observations in her discussion of Rosa Luxemburg's legacy to our challenges in the 21st Century:

> 'Luxemburg's argument about the pivotal role of militarism forces U.S. to ask how it functions as a sphere of accumulation (p. 77). Militarism cannot be divorced from economics, social upheaval, international solidarity, or environmental degradation. It is the key lens for understanding one of the state's fundamental roles in capitalism' (p. 80).

While Klein's *This Changes Everything* (*TCE*) is a very valuable contribution, especially as a critique of neo-liberal capitalism, it fails to identify and map out a plan to overcome the MIC and its imperial agenda, which prevents the establishment of a global cooperative regime to facilitate a rapid curb in carbon emissions and the creation of a wind/solar power infrastructure.

Klein briefly alludes to the imperial obstacle in *TCE* when she identifies the U.S. military as the biggest consumer of petroleum on the planet and points to the U.S. military budget as a potential source of revenue for a prevention program to avoid catastrophic climate change. But most critically, *TCE* fails to confront why the MIC, at the core of 21st Century capitalism, is such a huge roadblock to implementing a prevention program. Since global corporations include the main fossil fuel producers, in a world deriving 85% of its energy from this source, their carbon footprint is far greater than from direct military activities. According to the figures *TCE* cites for 2011, the total carbon dioxide equivalent emissions of the Department of Defense is less than 0.2% of the global total for that year (data from the International Energy Agency). While military expenditures are responsible for a colossal waste of energy and material resources that should be going to meet the needs of humans and ecosystems around the world (nearly US$2 trillion is spent annually on global military expenditures with the U.S. spending about half the total amount), even larger are the huge subsidies going to fossil fuels with indirect costs (including negative health impacts due to air pollution, as noted earlier, i.e., Coady *et al.*, 2015: US$5 trillion/year). With regard to the immense costs of the MIC, we also take note of Scheer (2007) when he pointed to nuclear and fossil subsidies as the greatest case of corporate welfare in world economic history (p. 131).

Furthermore, the nuclear industry is integrated into the MIC, and the threat of a nuclear attack is a long-standing instrument of imperial policy. The possibility of a nuclear war continues, with nuclear weapons now being possessed by nine nations, including notable hotspots such as Korea and the Middle East, with Israel possessing several hundred deliverable nuclear warheads. Recognizing Eisenhower's prophetic warning nearly sixty years ago, the MIC plays the dominant role in setting the domestic and foreign policy agendas of big capitalist powers, in particular the United States.

But most relevant to the threat of catastrophic climate change is the role of the Pentagon as the 'global oil-protection service' of the MIC (Klare, 2007). Scheer (2007) makes a similar point when he headed a section of his book with, 'The perversion of energy security by securing resources through military means' (p. 136). We ask the reader: would President George W. Bush have ordered the invasion of Iraq in 2003 if Iraq's main export was broccoli rather than oil? We submit that the U.S.'s imperial agenda actively blocks the global cooperation and equity required for a successful prevention program on climate change. The MIC is a continuing block to achieving global cooperation for a rapid curb on global greenhouse gas emissions and a full transition to wind/solar power. The Pentagon and NATO are part of the instrumental arm of the MIC's Imperial foreign policy, so the Pentagon going 'green' with respect to energy conservation and the use of renewables (Opalka, 2009) is simply a 'greenwashing' of its Imperial role. The Pentagon's recognition of the growing security threat from climate change, something that become highlighted in the context of the Trump Administration's climate denialism, reinforces the Imperial Agenda and military spending (see USA DoD, 2015). Notably, the U.S. Military's admission to the critical nature of the threat doesn't cancel out the contributions it itself makes to exacerbate it. Evidence of this can be observed in the fierce opposition to counting military-produced GHG emissions towards a nation's total emissions as measured in the context of international agreements, such as the UN Framework Convention on Climate Change (UNFCCC, 1992) and the Kyoto Protocol (1997). However, the recent Paris Agreement (October 2015) removes this exemption and includes military emissions for all nations as part of the emissions-cutting obligations set out for each country (Buxton, 2015).

Military spending and the MIC's Imperial Agenda are the critical obstacles posed by the MIC, not the sizable, but widely exaggerated greenhouse gas emissions of the Pentagon itself (McCollum *et al.*, 2009; Schwartzman, 2015). Critical contradictions within capital regarding the energy policy should be recognized, and the Green New Deal strategy must capture the 'solar' sector of capital in a multi-class alliance to force demilitarization and the termination of the perpetual war dynamic to have any hope of implementing a C3 prevention program in time.

This strategy remains very relevant: build a transnational movement for a Global Green New Deal (GGND). This is not a strategy that relies on the capitalist market driving 'green' capitalist investment, rather, it is one focused on opening up the path for a concrete C3 prevention program and a more favorable terrain for global systemic change (Schwartzman, 2011). Critical objectives of a GGND include the early nationalization of the energy industry in every nation, coupled with their decentralization, with community management and ownership of clean energy supplies, in a full transition to wind/solar power (Trade Unions for Energy Democracy, 2017). In Chapter 7 we discuss the GGND in more detail.

Paradoxically, the vast expansion of the MIC over the past half-century offers new opportunities for its dissolution, because humanity itself is now threatened on multiple fronts by its existence. There is a real possibility that continuing improvements in wind and solar technologies, resulting in dramatic increases in efficiency of capture and conversion, will undermine any remaining perception of the legitimacy of the MIC, thereby creating an upsurge of sufficient power to "finally put this Moloch back in hell, locking its gates of entry for eternity", reloading the poetic imagery of John Milton (1667, reprinted 2011) and Allen Ginsburg (1956).

The peace and climate/energy justice movements must be integrally linked; a shift to global demilitarization is a necessary condition for both robust cuts in carbon emissions and a transition to renewable energy on an adequate time scale. And, of course, demilitarization will open up the possibility of a massive reconversion of global production and consumption. What is most relevant to a global wind/solar transition is the vast resources, both material and human, that would be freed up in demilitarizing the global economy.

4.3.1. *What about the BRICS (Brazil, Russia, China and South Africa)?*

The BRICS nations are emerging as a potential economic and military counterforce to the Imperial Axis of the U.S./NATO, but with very contradictory aspects (Bond and Garcia, 2015). The BRICS global infrastructure program is in some respects even worse than that of its competitor (Alexander, 2014), particularly in its energy sector with its climatic impacts (note, e.g., China's plan to build more nuclear power plants and its role in helping to finance Hinkley Point). Nevertheless, we will make the following prediction, as crazy as it may sound: as both the highest carbon emitter and biggest investor in wind/solar technologies, China will emerge as the leader of a truly just and sustainable path for the rest of the World — this being driven by a class struggle with sufficient power growing out of both the huge negative impacts of its industrial infrastructure on its population and the paradox of its capitalist development under the banner of the 'communist' ideology, with remnants of 20th Century socialism still in place.

4.3.2. *How much energy will humanity need to confront energy poverty and C3 prevention?*

More attention should be focused on the message of the climate justice movement. For example, today's key players in the present climate justice movement present misleading prescriptions for change like the slogan, 'Keep the oil in the ground' (even worse, taken literally, 'Keep the oil in the soil'). This prescription ignores energy poverty, and presents an unrealistic framework for change. Instead the movement should argue first for a rapid phaseout of the highest carbon footprint fossil fuels (coal, natural gas and tar sands oil), the actual agenda of 350.org, using the *minimum* necessary amount of conventional liquid oil reserves to replace all fossil fuel consumption with a sufficient global wind/solar power infrastructure. Very welcome is the initiative to create Annex 0, coupled with COP21, to protect indigenous people from the extraction of hydrocarbons, but it too glosses over this issue (Oilwatch, 2015).

What will be sufficient wind/solar power to address humanity's and nature's needs? Short answer: a supply that is capable of terminating the energy poverty that now affects the majority of the World's people, while simultaneously facilitating climate adaptation, the sequestering of carbon from the atmosphere into the soil/crust, and bringing (and keeping) the atmospheric carbon dioxide level below 350 parts per million (ppm). Within the near future context, for an energy transition confronting the ever-mounting threat of catastrophic climate change, the following challenges point to the need for a significantly higher global energy supply than now, even after taking into account higher efficiencies possible in a wind/solar energy transition. With the present global consumption, a baseline (18 trillion watts of primary energy consumption), and growing global population, incremental energy will be required for the following new challenges facing our planetary civilization:

(1) Carbon sequestration from the atmosphere into the soil and crust to bring the atmospheric carbon dioxide level down, below the safe level of 350 ppm, and maintaining it below this level (the atmospheric carbon dioxide level is currently a bit above 400 ppm).
(2) The clean-up of the biosphere, notably toxic metals and other chemical and radioactive waste from the nuclear weapons, energy, and

chemical industries — a heritage of its long-term assault from the MIC, and other industrial wastes such as plastic particles in the ocean, threatening its ecosystems (Harrabin, 2017a).

(3) The repair and expansion of physical infrastructure, such as electrified rails, and the creation of green cities globally.

(4) Adaptation to ongoing climate change, especially by the global South with its disproportionate impacts, even if warming could be kept to below 1.5°C.

All four imperatives will require very significant energy supplies from future wind/solar power, incremental to the present consumption level. The actual level of this increment needs study but some preliminary estimates are now available. For example, if 100 billion metric tons of carbon, equivalent to 47 ppm of atmospheric CO_2, were industrially sequestered from the atmosphere it would require 5.9 to 18 years of the present global energy delivery (18 TW), assuming an energy requirement of 400 to 1200 KJ/mole CO_2 utilizing a solar-powered high efficiency source of energy (Zeman, 2007; House *et al.*, 2011). This requirement would of course be reduced by the use of agriculturally-driven carbon sequestration into the soil (see P. Schwartzman and D. Schwartzman, 2011; D. Schwartzman and P. Schwartzman, 2013).[4.3]

[4.3] With regard to the required energy for Carbon Dioxide Removal (CDR) from the atmosphere we made following estimate: from their estimated energy required per mole of CO_2 sequestered, to reach 350 ppm CO_2 in atmosphere by 2100, an equivalent of 328 ppm must be sequestered, equal to 695 Pg C assuming a constant rate (see Hansen *et al.*, 2017, Figure 10b). According to Smith *et al.* (2016), this would require an increment to the global energy supply of an equivalent of 16.7 TW (80 years of sequestration, starting in 2020, requiring 527 EJ/year; note this assumes that 1 mole of CO_2 requires 727 KJ for sequestration, near the upper limit of 1200 KJ assumed in D. Schwartzman and P. Schwartzman, 2013, and in Correction available on www.solarutopia.org). The 6% decrease/year in CO_2 emissions scenario over the same 80-year period would correspond to sequestration of 153 Pg C, and an expenditure of energy corresponding to an average of 3.66 TW per year. Coupling the DAC sequestration with agroecology and reforestation — both which require significantly less energy — would of course bring down the required energy expenditure. Nevertheless, a significant incremental energy supply to the present is imperative to have any chance of reaching the 350 ppm CO_2 goal by 2100, and because of continued ocean equilibration with the atmosphere, sequestration into the crust and soil must continue after 2100.

Carbon sequestration from the atmosphere will require a very ambitious program involving a combination of technologies, including the transformation of agriculture to agroecologies, as well as reacting carbon dioxide and water with mafic rocks and crust to produce carbonates; this is not 'clean coal', i.e., carbon capture and storage (CCS) from coal-derived emissions. The following conclusion of a recent paper is very relevant:

> 'We conclude that CDR [Carbon Dioxide Removal from the atmosphere] can be a game changer for climate policy in the sense that it significantly improves feasibility and cost considerations for achieving stringent climate stabilization. It is, however, a **complement, not a substitute to the traditional approach of mitigating emissions at their source'** (Kriegler *et al.*, 2013, p. 55).

This sequestration program will be imperative for the rest of this century and beyond because approximately half of the anthropogenic (caused by humans) carbon dioxide emissions go into the ocean and biota, which continuously re-equilibrate with the atmosphere (Cao and Caldeira, 2010; Gasser *et al.*, 2015). CDR is a technology that urgently needs more research and development to have any chance of limiting warming to 2°C, or the better 1.5°C (Van Vuuren *et al.*, 2017).

Natural climate solutions, i.e., soil conservation, restoration, and/or improved land management actions can potentially contribute to significant climate mitigation by increasing storage of soil carbon (Griscoma *et al.*, 2017) but more aggressive measures will be needed to limit warming to no more than 1.5°C. Efforts to boost sustainable agriculture, specifically with agroecologies and permaculture, are imperative to replacing industrial/GMO agriculture, both to confront the challenge of climate change and to eliminate big negatives of the present system of unsustainable agriculture. These alternatives will be very useful in sequestering carbon from the atmosphere, burying it in the soil. Hence, we applaud the launching at COP21 of the Soils for Food Security and Climate (LPAA, 2015). But some have even claimed that a transition to sustainable agriculture alone can reverse global warming without the elimination of GHG emissions from fossil fuel sources (e.g., according to Biodiversity for a Livable Climate (2015), 'We can return to pre-industrial atmospheric carbon levels in a few decades or less, and cool the biosphere even faster than that'). Kastner (2016) makes similar arguments. The critical issue is the potential global sequestration flux assuming a complete transition to sustainable agriculture. The maximum flux is far too small to achieve what is claimed, even if fossil fuel emissions cease immediately (Schwartzman, 2015c).[4.4]

Further, as pointed out in Schwartzman (2015c), ongoing warming will plausibly reduce the potential of this mode of carbon sequestration owing to the enhanced respiration of soil carbon back into atmosphere as temperatures increase as a result of the decomposition rate of this carbon pool. Indeed, Crowther *et al.* (2016) found strong evidence for this impact of climatic warming, likewise Melillo *et al.* (2017), see commentary by Metcalfe (2017). In addition, based on a recent study of radiocarbon age, the soil carbon sequestration potential may have been significantly overestimated (He *et al.*, 2016). In conclusion, to achieve the goal of limiting warming to below 1.5°C by 2100, carbon-sequestration from the atmosphere technologies, in particular with atmospheric carbon going into the crust, will be imperative, instead of simply relying on carbon-sequestration into the soil. One very promising approach is injecting atmospheric carbon dioxide into basalt rock (Oelkers *et al.*, 2016); see Schwartzman (2016b) for more on this subject. Carbon sequestration from the atmosphere into the soil and crust is the only imperative climate geoengineering approach. The dangers of other climate geoengineering technofixes have been highlighted in many publications (e.g., Wetter and Zundel, 2017). In particular, the threat to biodiversity from sudden implementation and termination has been recently modeled (Trisos *et al.*, 2018).

With respect to the need for climatic adaptation, note that a recent study finds 'abrupt changes in sea ice, oceanic flows, land ice, and terrestrial ecosystem response, although with little consistency among the models. A particularly large number is projected for warming levels below 2°[C]' (Drijfhout *et al.*, 2015, p. E5777). The costs of climate

---

[4.4] Hudak (2017) made a systematic critique of the claimed potential of livestock grazing as an enhancer of soil carbon storage.

adaption are already near US$50 billion per year and are expected to rise exponentially with ongoing climate change, especially in developing countries (Dougherty-Choux, 2015).

A greater energy capacity than present will likely be required to realize these objectives. Even with a complete transition to a global wind/solar power infrastructure by 2050, including its roughly 30% gain in efficiency, at least 22 trillion Watts will be required to guarantee the minimum energy per person necessary for a state-of-the-science quality of life (3.5 kW/person × 0.7 × 9 billion people; note that the latest United Nations (2015) projection for 2050 is 9.7 billion people). Additional capacity will be necessary especially for ongoing carbon-sequestration from the atmosphere and climatic adaptation, with the total required likely approaching the order of 25 trillion Watts. This approach is imperative because these applications will require extra capacity being created as early as possible, given the physics of greenhouse gas forcing (Matthews and Caldeira, 2008).

4.4. Can energy efficiency radically lower the energy required for the OWSP?

The importance of increasing energy efficiency in a global shift to a renewable power infrastructure is well recognized (IRENA, 2017). Cullen *et al.* (2010; 2011) argue for a potential radical increase in energy efficiency, with 85% of global energy use being saved by theoretically achievable efficiencies in passive and conversion systems (most from passive systems). However, since this is a theoretical potential, Sorrell (2015) argues that the technical and especially economic potential is likely much less, pointing out that the 'rate at which improvements in conversion efficiency or passive systems can be achieved is constrained by the rate of turnover of the relevant capital stock', with the example of cars having an average lifetime of 10 years, and the lifetime of power stations, blast furnaces, ships and aircraft easily exceeding 30 years (Davis *et al.*, 2010). Further, according to Sorrell (2015), the premature replacement of existing equipment can accelerate the rate of efficiency improvement, involving trade-offs between the energy used in constructing new equipment and the energy used in operating the old equipment (Lenski *et al.*, 2010).

Von Weiszacher *et al.* (2009) made a case for radical improvements, up to 80%, in resource productivity using state-of-the-science technologies. Likewise, Krausmann *et al.* (2017) emphasized the potential for increasing the current low recycling rate of material stocks (12%):

> 'Reducing expected future increases of material and energy demand and greenhouse gas emissions will require decoupling of services from the stocks and flows of materials through, for example, more intensive utilization of existing stocks, longer service lifetimes, and more efficient design.'

The development of new technologies will undoubtedly increase the energy efficiency of work done for societal needs, such as air conditioning (e.g., Hanley, 2018). Jacobson (2017a, b) has argued for a 30% energy efficiency gain with solarization. We have little doubt that even bigger efficiency gains are ultimately achievable. But most of the people in the World now need more energy, notably in China, India, and Africa, so if Cullen *et al.*'s theoretical efficiency were somehow applied now, energy poverty would remain. If this efficiency were applied to the U.S. alone, the kW/person would go from 9.06 to 1.36 (or 2.27 for 75% gain in efficiency). So there are at least four sets of countries as classified by the coupling of energy efficiency and life expectancy, see Table 4.2 below; note that countries in group (4) suffer most grievously from energy poverty, while countries in groups (2) and (3) can benefit most from increases in energy efficiency.

For a future world of 9 billion with 2 kW/person enjoying clean energy and state of the science life expectancy, the total primary energy consumption would be 18 TW, the same as now, but not counting the ongoing incremental needs already noted and discussed. So, we conclude that the World needs more energy than now, at least until a truly sustainable steady-state global economy is achieved.

4.5. The great transition

Returning to the issue of future human population forecasts, we suggest that the biggest uncertainty in these projections ahead in the next few

Table 4.2. Countries classified by life expectancy

	Life Expectancy (at birth, yrs)	Power per capita (kW/person)
(1) Already energy efficient countries (but could be even more efficient) with high life expectancies		
Japan	83.7	4.6
Spain	82.8	3.3
Italy	82.7	3.2
UK	81.2	3.7
Portugal	81.1	2.7
(2) High life expectancy, but higher energy consumption per capita		
Australia	82.8	7.1
Sweden	82.4	6.6
S. Korea	82.3	7.0
Canada	82.2	10.5
Norway	81.8	7.4
(3) Energy wasteful countries (lower life expectancies, high energy consumption per capita)		
U.S.	79.3	9.2
Saudi Arabia	74.5	9.2
Russia	70.5	6.6
South Africa	62.9	3.6
(4) Lower life expectancies, lower energy consumption per capita includes energy wasteful countries like China (high energy intensity as measured by primary consumption/GNP, and those with very low energy consumption per capita (labor intensive production, most of Africa)		
China	76.1	3.0
India	68.3	0.8
Mexico	76.7	2.0
Brazil	75.0	2.0
Nigeria	54.5	1.0
Egypt	70.9	1.1
Indonesia	69.1	1.2

Note: For these countries, climate is an obvious factor in raising energy consumption, heat in the winter or air conditioning in the summer (Africa).

Sources: Life Expectancy: WHO (2016), 2015 data.
Power per capita: World Bank (2014), 2014 data.

decades is contingent on how bad will the impacts of dangerous climate change be, and to what extent will humanity succeed in a prevention program. If humanity fails, then climate hell with potential for mass human die-off; if humanity succeeds, then population stabilization. Some climate justice activists accept the collapse of civilization and call for a radical reduction in global energy consumption regardless of its impact on humanity. For example, Derrick Jensen (2012), an extreme anti-extractionist, calls for an immediate shut-down of all oil wells (taking 'keep the oil in the ground' literally). This strategy is extremely problematic. Its implementation would prevent a solar transition with the capacity to both eliminate energy poverty and work through the climate crisis. Only a global clean energy infrastructure supplying more energy than is produced now has the capacity needed for ongoing climate adaptation and the sequestering of carbon from the atmosphere into the soil and crust as components of a catastrophic climate change prevention program. Of course, realizing this goal poses critical political challenges (Fell, 2017).

A global wind/solar transition replacing present unsustainable energy supplies must be parasitic on these supplies, just as the industrial fossil fuel revolution was parasitic on biomass (plant) energy, until it replaced the former supply with sufficient capacity. Liquid oil, which has the lowest carbon footprint of the fossil fuels, is the preferred energy source to make this renewable energy transition.

With the creation of a cooperative global regime on climate change, each nation will have an opportunity to fully benefit from this transition, while contributing resources compatible with their naturally existing oil, wind, or solar resources. Thus, oil-rich countries (e.g., Venezuela (Schwartzman and Saul, 2015) and countries in the Middle East) will be valuable partners in this transition, which is only possible by creating a cooperative global regime, with the dissolution of the Military Industrial Complex ('MIC') and its imperial agenda at the core of 21st Century Capitalism.

4.5.1. *Stranded assets of fossil fuels and their implications*

The issue of stranded assets of fossil fuels is now looming as a critical challenge to the profitability — indeed, the very survival — of the corporate/banking sectors owning and financing these assets, as well as to the widening movement to divest from fossil fuels. 'Stranded assets of fossil fuels' refer to the potentially lost deposits still in the crust (not the soil, despite the welcome motivation of those who say 'keep the oil in the soil'!), in resources owned primarily by transnational petroleum companies such as Exxon-Mobil, Royal Dutch Shell, as well as state entities (e.g., Saudi Aramco, China National Petroleum Corporation). Of course, what is driving this potential is the fact that a large fraction of these fossil fuel reserves must stay in the crust, and be 'unburnable' to have any remaining chance of keeping future warming below the goals set by the Paris Agreement (2°C; though it is much better to keep it below 1.5°C).

Estimates of the still burnable fossil fuels to keep temperatures below 2°C correspond to no more than one-fifth the total reserves (Carbon Tracker, 2011; 2015), in particular, one-third of oil, one-half of natural gas and only one-fifth of coal reserves (McGlade and Ekins, 2015). The limit on natural gas reserves is likely even smaller, as a result of methane leakage into the atmosphere (Hendrick *et al.*, 2016). Obviously, to keep below the 1.5°C warming limit, even a smaller fraction is still burnable (e.g., Knopf *et al.*, 2017; Rockstrom *et al.*, 2017).

Potential stranded asset losses have been estimated at US$2 trillion (Climate Tracker, 2015), while Citigroup gives a much larger US$100 trillion figure in anticipation of COP21 (Parkinson, 2015). This challenge is now seriously recognized in the climate change literature (e.g., Fuss, 2016). But equally or even more significant is the impact of ongoing climate change on financial assets (Dietz *et al.*, 2016; Battiston *et al.*, 2017). The synergism of both relationships clearly has the potential of undermining the stability of the global capitalist system. The World Wildlife Foundation's Lorenzo (2016) shows how serious this prospect is with regard to the financial resources necessary to keep global warming below the 1.5°C limit. Nafeez Ahmed (2016) cites Pfeiffer *et al.* (2016) for pointing out that the threat of stranded assets includes not just unburnable fossil fuel reserves, but also the huge global carbon-intensive electricity infrastructure dependent on fossil fuel energy.

Both Richard Heinberg (2016), Senior Fellow at the Post Carbon Institute, and Ahmed (2016) emphasize the threat of the collapse of the oil industry driven by declining EROEI ratios of this fossil fuel. But we find their analysis confusing since it conflates physical with political economic constraints now in place. We should welcome 'peak oil', even better 'peak coal, tar sands and natural gas', with all three being the highest carbon footprint fossil fuels, peak here referring to a peak in global production, the sooner the better for any chance for climate security in a robust energy transition to a global wind and solar infrastructure coupled with complete phaseout of fossil fuel consumption. Both Heinberg and Ahmed claim the energy highway is crumbling beneath us as the profitability of oil production declines, but we argue that this is a political economic outcome, not a physical one. The low price of petroleum in the period from mid 2014 through 2017 appears to be the direct result of US–Saudi collusion. The U.S. imperial goal of regime change in Venezuela apparently converged with, as Heinberg points out, the Saudi objective of driving out competition from shale oil derived by fracking. Hence the Saudis blocked effective action by OPEC to limit petroleum production, thereby depressing the price of oil. Both Heinberg and Ahmed cite the low EROEI of bitumen oil but this is irrelevant to the issue of the production of higher EROEI conventional oil being a bridge fuel in a rapid wind/solar energy transition, especially since the EROEI ratio of these technologies continues to increase, hence requiring less and less existing energy for their creation. Heinberg is correct to point out the environmental risks involved in even conventional oil production, so a strong regime of protection is required, while extracting the minimum amount of oil necessary to power a complete global transition to wind/solar energy supplies. A higher price for oil, including carbon taxation, would be beneficial in multiple ways, including increasing investment in wind/solar. This price increase should be coupled with compensation to low-income commuters and consumers of oil for home use until adequate low cost to free public transport is provided along with other alternatives to oil consumption.

Yes, as Heinberg emphasizes, oil is now essential for certain uses, in particular for transportation and shipping, but this can and should be replaced in transition to wind/solar power by both electrification, especially for rail/public transport, and producing carbon-neutral hydrocarbon fuel (particularly for air travel) derived from solar-powered chemical reactions using carbon dioxide and water (e.g., Asadi *et al.*, 2016).

Heinberg (2016) ignores the studies of Mark Jacobson and collaborators (e.g., Jacobson and Delucchi, 2011; Delucchi and Jacobson, 2011) which demonstrate that a renewable energy transition is plausible in a few decades. Only a fraction of the global reserves of conventional liquid oil is sufficient to power this energy transition, while keeping the temperature increase below 1.5°C by 2100, the goal of the Paris Agreement — but only if this transition starts very soon with the highest carbon footprint fossil fuels phasing out rapidly (Hansen *et al.*, 2013, 2017; also modeled in P. Schwartzman and D. Schwartzman, 2011 and Schwartzman *et al.*, 2016).

As Knuth (2017) so effectively emphasizes, the challenge of fossil fuel stranded assets should be confronted as an unprecedented opportunity to de-commodify this energy source, a process of 'green enclosure' — the opposite of the enclosure practiced in the birth of both capitalist agriculture and industry and the continuing land grabs of transnational corporations, particularly in the global South. Knuth points to the fossil fuel divestment movement, a major project of 350.org (see Rowe *et al.*, 2016), leading to the goal of the 'moral obsolescence' of the fossil fuel industry, thereby serving the 'intra-capitalist struggle' so critical to the unfolding of a truly GGND. Thus, the growing global movement for fossil fuel divestment (Fortuna, 2017) should be seen as a component of the ecosocialist class struggle for strengthening the 'environmental regime' (Sandler, 1994).

4.5.2. *Green capitalism builds big solar: Should we throw the baby out with the bathwater?*

A recent headline reported: 'Wind, solar power soaring in spite of bargain prices for fossil fuels' (Warrick, 2016). Should we welcome this trend, or focus only on the fact that capitalist-driven solarization is fraught with problems?

For example, Hamouchene (2016) provides a very thoughtful and thorough critique of how the capitalist-driven creation of renewable energy infrastructure is lacking social management and planning. Further, the article points out that while such large projects in the global South should prioritize the elimination of energy poverty, this is not apparently the plan for Morocco and North Africa. These challenges should thus be on the front burner for ecosocialist intervention in ongoing class struggles during the energy transition.

In an earlier article, Hamouchene (2015) critiqued an even bigger solar project on the drawing boards, Desertec, on precisely the grounds the senior author warned twenty years ago in Schwartzman (1996) where he noted: 'plausible scenarios of continued neocolonial subjugation of the "south" under the rubric of promoting solar energy are conceivable (e.g., a Saharan photovoltaic network controlled by transnationals supplying power to Europe under highly unequal arrangements of exchange)', although the actual plan has been to build a huge concentrated solar power infrastructure much bigger than the Ouarzazate Solar Plant in Morocco, rather than photovoltaics in the Sahara. Such a project could potentially supply the current global electricity consumption on less than 6% of the Saharan land area.

Nevertheless, we have some disagreements with Hamouchene regarding the following statements in the Ouarzazate article:

> 'One needs to say it clearly from the start: the climate crisis we are currently facing is not attributable to fossil fuels per se, but rather to their unsustainable and destructive use in order to fuel the capitalist machine. In other words, capitalism is the culprit, and if we are serious in our endeavors to tackle the climate crisis (only one facet of the multi-dimensional crisis of capitalism), we cannot elude questions of radically changing our ways of producing and distributing things, our consumption patterns and fundamental issues of equity and justice' (Hamouchene, 2016).

While we agree with the last sentence, the first is problematic. Surely, we cannot replay the history of fossil fuel consumption driven by fossil capital. Carbon emissions come from the actual burning of fossil fuel, a physical process. This is the prime driver of the climate crisis, to be sure, a result of the historical use of this fuel as the energy source in the reproduction of capital. We cannot imagine what would have been the alternative to fossil fuels' 'unsustainable and destructive use' *while still consuming fossil fuels,* except in a rapid transition away from these fuels in a solar transition. Further, Hamouchene says:

> 'It follows from this that a mere shift from fossil fuels to renewable energy, while remaining in the capitalist framework of commodifying and privatizing nature for the profits of the few, will not solve the problem. In fact, if we continue down this path we will only end up exacerbating, or creating another set of problems, around issues of ownership of land and natural resources.'

Well if this 'mere shift' can be accelerated, as climate science tells us is imperative to have any remaining chance to avoid climate catastrophe, with the majority of humanity living in the global South bearing the heaviest impact, then such a shift to renewable energy could have a stupendously positive result: the end of energy poverty in the global South and the capacity to bring down the atmospheric carbon dioxide level below the safe upper limit of 350 ppm, preventing climate catastrophe! And yes, *the movement for climate and energy justice should not accept this transition while remaining in the capitalist framework. On the contrary, we should use this unprecedented opportunity to move to a post-capitalist future*.

We are now confronting a clean energy transition that is still too slow. And only when a more robust renewable creation is coupled with the rapid phase-out of fossil fuels — starting with the fossil fuels with the largest carbon footprints (i.e., coal and natural gas because of methane leakage to the atmosphere) and tar sands oil — will there be any chance of avoiding climate catastrophe (to be discussed further in Chapter 8, Section 8.5). And there is every reason to believe that a *full transition with these characteristics* cannot be generated in the capitalist framework, given the dominant role of the Military Industrial Fossil Fuel Complex. While the ecosocialist class struggle is still too weak to prevent the deficiencies in these big solar projects, as the global climate and energy justice movement gains strength the opportunity to create a sustainable and just solar transition will grow. Given known environmental and labor hazards associated with the solar industry, e.g., modern forms of silicon PV production (OSHA (2018), this transition should include a robust protection regime for ecology, workers and the community.

But the creation of a wind/solar energy infrastructure should be welcomed now even with all the problems pointed out by Hamouchene. We cannot wait for the end of the rule of capital to start building this renewable infrastructure; it will be too late.

To simply say capitalism must be replaced by socialism is a conclusion, not a strategy, and to claim that capitalism must be first replaced by socialism to prevent catastrophic climate change is a cowardly rejection of responsibility to living and future generations. However, to simply rely on green capitalism to implement a prevention program using the usual market-driven mechanisms is a recipe for disaster (e.g., see critiques by Sweeney, 2015; Wright and Nyberg, 2018). Global greenhouse emissions continue to climb because market-led renewable energy growth is much too slow to replace fossil fuels in the short window of time remaining, and fierce corporate resistance continues to block the replacement of industrial/GMO agriculture by agroecologies.

In *TCE*, Klein outlines the radical reforms necessary to avoid climate catastrophe, including components of a Global Green New Deal (GGND), a UN proposal from 2009. This is a good start. But she fails to name a real alternative to unsustainable capitalism — ecosocialism, the only viable socialism of the 21st Century. Nor does *TCE* provide a concrete vision of the 'other world that is possible' after capitalism is eliminated on our planet; neither ecosocialism nor its leading thinkers are even listed in the index. Perhaps Klein's omission is strategic — that she's fearful of stepping too far ahead of popular consciousness, even within the broad climate justice movement.

Moodliar (2015), who wrote one of the most perceptive reviews of *TCE,* argues that Klein leaves out the importance of class struggle. While the book doesn't engage much with Marxist theory, we disagree with Moodliar's assessment. Klein vividly describes the myriad climate justice movements whose struggles constitute class struggle — multi-dimensional and transnational — even if the global political Subject they encompass is not yet fully conscious of itself. Only when this consciousness emerges will we have a chance to avoid climate catastrophe. But there are very encouraging signs that this Subject is being born, e.g., the Leap Manifesto from Canada (The Leap Manifesto, 2017).

Critics of the neo-liberal worldview commonly call for an end to economic growth as a necessary condition for ecological sustainability and for any chance to prevent C3 (e.g., Trainer, 2011; Anderson, 2015). To be sure 'business as usual' growth will lead humanity to C3. But the qualitative aspects of economic growth are critical in confronting the immediate challenges posed by the threat of C3 as well as the undeniable lack of material consumption enjoyed by the majority of humanity living in the global South, most critically, their state of energy poverty (Schwartzman, 2014b; 2014c). We revisit this issue in more detail in Chapter 7.

4.6. Conclusion, why *solar*?

Solar is by far the most abundant source of energy and the technologies to harness it are already in use. And, given a robust social management process during its lifecycle, solar also has very low negative health and ecological impacts. Moreover, a global transition to solar is actually achievable in the available time frame to avoid catastrophic climate change, and under real existing capitalism, solar is the energy source most compatible with decentralized, democratic management and control, and relatively free of the dictates of the MIC compared to fossil fuels and nuclear power. The dissolution of the MIC and its conversion to a sustainable physical and political economy is simultaneously a requirement for preventing catastrophic climate change and removing a major blockage to the other world that is possible. Let us dare to make this a reality while there is an ever-diminishing window of opportunity.

Chapter 5

A New Food Future — Agroecology

5.1. Introduction

As indicated in previous chapters, it is imperative that everyone in the world is fed without harm or any reduction in their ability to feed their descendants. Additionally, agriculture must be carried out in a way that doesn't destroy the planet's other species' ability to thrive as well. Given agricultural history over the past ~80 years and the directionality of many national and international programs, we clearly have great challenges ahead of us. Fortunately, there are alternatives that work and expansion of these will have incredible benefits for humanity and the entire biosphere.

5.2. Our dominant food paradigm

As we have seen, current dominant modes of agriculture focus on the monopolization and maximization of profits through various means. Kirschenmann (2004) highlights the centrality of increasing yields:

> 'The emergence of this exclusive money-based, laissez-faire, libertarian economic theory, and subsequent economic practices, led to a conviction in agriculture, as well as in all other human enterprises, that the market always knows best and that price alone captures appropriate value.'

This productionist ethic, Kirschenmann adds, is morally bankrupt and allows for many types of reckless behavior. Biel (2016) highlights the imperialistic characteristics of modern agriculture:

> '[The Green Revolution (GR) programme promoted] hybridized "high-yielding varieties" (HYVs) of rice and wheat…deliberately bred so that they would only function with high inputs of chemicals … and machinery manufactured by the corporations which sponsored the GR' (pp. 75–76).

While we have seen increases in yields of some 20% in recent years (for 174 crops tracked by the UN, from 1985–2005), largely due to technological advances, these increases have come at a steep price to the environment (Foley *et al.*, 2011). The two values — monopolization and profit — driving modern industrial agriculture are inherently at odds with human systems and, particularly grotesque when applied to agriculture. Monopolization fails high quality food production because of the cultural differences that fortunately still exist within human society — as yet, not everyone chooses to drink corn syrup-laden Coke and eat GMO-corn generated hamburgers. It fails as well because of the imbalance of power and influence that comes with monopolistic enterprises. Reduced diversity in business practices is inherently less resilient and more unstable as it would be in ecological systems. Maximizing profits results in focusing on sales rather than consumer health, livable worker wages, animal health, and environmental damage. As a result, huge (and continued) negative costs and suffering have occurred in all human and life systems. KPMG International's examination of 22 environmental impacts of 11 key sectors of society establishes food production's sizable footprint to be larger than all other ten sectors, including gas and oil production (KPMG, 2012). Other social values should take precedence if humanity and nature is to thrive.

Yet, a simple resetting of values has not taken root at the rate it needs to. Dominant paradigms are hard to change and industrial agriculture is certainly one that will require a major overhaul. Simply put, many (if not most) of those in power globally are of the persuasion that intensive agriculture of 'advanced' societies produces food more efficiently and in higher quantities than any other food system ever created. Consequently, floods of propaganda fill U.S. newspapers, scientific journals, and national and international organizations (such as the Farm Bureau and the World Bank). Given this dominant discourse, it would seem asinine to move against such a system as modern agriculture. As the story goes, the Green Revolution (of the 1960s and 1970s), driven by the introduction of new, advanced technologies, saved millions of lives and almost eliminated starvation. And as long as we continue to introduce these types of technologies (using the latest and greatest biotechnology has to offer), all will benefit and health will improve. These arguments are pervasive and deserve a closer look.

Expanding on our discussion in chapter 3, let us look at and question the three most relevant boasts promulgated regarding modern agriculture. Once we shed doubt on their veracity, we will be ready to explore new and alternative paradigms of food production that will provide sustainable solutions to one of our greatest challenges. Here are a few of the most prominent and relevant claims/boasts:

(1) Big corporate agriculture is the best form of agriculture. This sentiment is captured in the following quote: 'large farmers — who are responsible for 80% of the food sales in the United States, though they make up fewer than 8% of all farms — ...are among the most progressive, technologically savvy growers on the planet. Their technology has helped make them far gentler on the environment than at any time in history. And a new wave of innovation makes them more sustainable still' (Lusk, 2016; Dr. Lusk is a professor of agricultural economics at Oklahoma State University).

(2) Genetic engineering is the key to feeding all humans. This sentiment is captured in Prakash (2000), which claims that 'biotechnology is one of the best hopes for solving the food needs of the poor today ... and certainly in the next 30 to 50 years, when there will be 9 billion on the globe'.

(3) Modifying the genetic makeup of organisms is safe. This claim is captured in Prakash (2000), who believes 'products from biotechnology are no less safe than traditionally bred crops. In fact, they may be even safer, because they represent small, precise alterations with the introduction of genes whose biology is well understood'.

These claims serve to promote an agricultural future that is dominated by ever increasingly large farms (in the U.S. and abroad — so that, by necessity, there are fewer and fewer farms), more and more GMOs and other biotechnical solutions, and the increased consumption of processed foods and meat. Thus, if these claims are false or, at least, highly questionable, and they are, then it raises the question regarding the future of food, particularly the dominant industrial methods and techniques currently used to produce it.

The environmental problems reaped by industrial farming are legendary, so much so that Dr. Lusk's statement must have been parody (unfortunately, it wasn't meant to be). Expanding on evidence provided in Chapter 3, the 'Dead Zone' in the Gulf of Mexico around the mouth of the Mississippi has often exceeded 8,000 square miles since 2000 (this is about the size of New Jersey). Driven by excessive nutrient loads that have been dumped on Midwestern grain and soybean fields to enhance soil fertility, these zones represent a disaster that continues to get worse not better (Philpott, 2016). Increasing areas of the Great Lakes and the Chesapeake Bay are now also exhibiting oxygen-deprived areas (Milman, 2017). Synthetic pesticides (which include herbicides, insecticides, rodenticides and fungicides) have poisoned the biosphere in incalculable ways (Levin, 2007; Steingraber, 2010; Dowdall and Klotz, 2016). First generation synthetic chemicals were used as warfare agents for WWI and WWII. Derivatives of these chemicals have been sprayed/applied (and therefore have drifted, leached, and/or transferred) across the planet in unimaginable quantities for the past 90 plus years (Steingraber, 2010). Consider that 1.1 billion pounds of active ingredient biocides were used in the U.S. alone, during 2006–2007 — that is more than 3 pounds per U.S. citizen (Grube *et al.* 2011). And since, compositionally, pesticides are 99% 'inert' ingredients (examples include, 'emulsifiers, solvents, carriers, aerosol propellants, fragrances, and dyes'; EPA, 2017) and, since 'inert' does not guarantee safety (as most of these chemicals have not been properly tested), the actual dosage on the landscape is probably several orders of magnitude higher. Since most pesticides used today are broad spectrum (meaning they kill or detrimentally affect the reproduction of a great variety of organisms), they often destroy beneficial plants and insects as well as 'agriculturally-problematic' (yield-reducing) ones. The workers that apply these chemicals aren't immune to their toxicity either. Annually, the use of these chemicals results in an estimated 250,000+ deaths and three million poisonings worldwide (WHO, 2004). Add to these problems, the fossil fuel addiction of modern agriculture (in its production of fertilizers, pesticides, hormones, and antibiotics) and the causal impacts of catastrophic climate change (C3), as well as modern large scale industrial agriculture clearly does not lead the World in a promising direction.

Claims (2) and (3) above concern biotechnology while emphasizing the efficacy of genetically modified organisms (GMOs). This perspective received a major boost recently when a widely-publicized research study conducted by the National Academy of Sciences (NAS) (2016) found 'no conclusive evidence of cause-and-effect relationships between GE [genetically-engineered] crops and environmental problems', Furthermore, by comparing the incidence of specific health problems in the United States (where GMO food has been widely available since the 1990s) to those found in the United Kingdom and western Europe (where GMO foods have not been widely consumed), the study found no evidence that GMO foods are causing any adverse impacts on humans, in spite of the continuing disparities in the life expectancy between the U.S. and European nations. So, is the safety debate 'over' as so many, for example, according to climate writer Mark Lynas (2016), have been eager to point out? Not so fast.

While biotechnological knowledge may have a lot to offer us in the future, GMO foods cause problems, have unknown risks, don't live up to their promises and are flatly unnecessary. Swanson *et al.* (2014) found significant health problems in the U.S. strongly associated with the use of chemicals commonly dispersed on GMO corn and soybeans. Modern agriculture's increased reliance on GMO seeds presents the serious problem of reduced genetic diversity globally. As of 2014, about half of all farmland in the U.S. were planted with GMO crops. Even the U.S. Dept. of Agriculture admitted in a 2014 report (now, no longer easily obtained on its website), that there were issues with this. Growing millions of acres of monocrops of soy, corn, wheat and an increasing number of other proprietary crops reduces biodiversity by definition and increase the possibility of a major widespread contagion (and subsequent crop failure) but they also inject their 'foreign' genes into native plants through pollination and create genetic modifications beyond the initial GMO crop. GMO crops have their genes drift, resulting in 'adventitious presence', the unintentional mixing of genes, which makes it nearly impossible for farmers who do not want GMO genes in their crops. Given the dominance of GMO seed injection into the global environment, it is safe to say that we are engaged in a massive experiment that we do not know the outcome of and that we cannot likely stop at this point (Arcieri, 2016). The precautionary principle would clearly, and emphatically, reject our current course for this reason alone.

Beyond biodiversity concerns, GMO crops very often are paired with an herbicide or insecticide, some of which have been shown to be 'probable' carcinogens when careful examination is carried out. That was the WHO's determination of glyphosate, the main ingredient in Monsanto's Roundup, a widely used herbicide (Cressey, 2015). However, we do not deny that in 2017, the UN and WHO seemingly switched their position, specifying that this chemical unlikely posed a threat through diet, though it remained uncertain about the dangers of occupational exposure (of farmers in the fields, for instance) (Neslen, 2016). And, even as the NAS study admits, we are just beginning the era of GMO seeds, and therefore, we must acknowledge that once these genes enter into the World's ecosystems, they cannot be retrieved at some later time when dangers become more apparent; this situation is not so dissimilar to lead contamination in the environment which will continue to wreak havoc for generations to come, long after this toxic substance was finally removed as an additive in gasoline and paint.

Additionally, let us remember the two key promises made by GMO producing companies: that GMO seeds would (a) increase yields, and, (b) decrease toxic chemical use. Both these claims have been seriously challenged. Since the introduction of GMO seeds (for crops including corn, cotton and soybeans), herbicide use is up over 20% and most GMO crops don't show any increased production when compared to European varieties that aren't GMO (Hakim, 2016); for a noteworthy recent exception, see the meta-analysis done on corn by Pellegrino *et al.* (2018). Continuing along our current GMO path should certainly be abandoned if its two core reasons for advancement and implementation of them are found to be spurious.

And while there are many more reasons to be apprehensive about GMOs (such as the lax governmental regulations of them in the U.S., industry's questionable 'safe claim' track record, and industry's centralized monopoly on seeds and therefore food), the main reasons to avoid them are because they are not necessary and they do not promote sustainable practices. For these two reasons, which are often flatly ignored by proponents of GMOs, we should be moving in a different direction than advocated by corporate-dominated agriculture and many organizations. Fortunately, as we will present now, there are alternative methods that exist without the inherent risks and loss of autonomy found in modern industrial agriculture with its emphasis on synthetic chemicals and GMOs (Arcieri, 2016). When one considers the full options available to humanity and does not just lock in with the "status quo" (and the billions and billions of dollars that promote it), we think rational people will recognize that another, and better, world is possible and necessary in the arena of agriculture and food.

5.3. A new paradigm — a Just Green Revolution through agroecology

If we were to 'start over' and develop an agriculture/food system that met the goals of humanity, what would it look like? (We actually do not need to start over but it helps to think with a clear slate in order to clarify one's true aspirations and goals.) First and foremost, we would create a food system that could maintain itself (and the human population) in perpetuity — sustainability in its most extensive form. Undoubtedly, it would need to produce more food than we do now, as populations continue to grow and rises in affluence demand more meat products per person. Imagine if Chinese people begin eating meat at the rate of Americans. They would need to import huge amounts of feed for the animals they were raising. Changes of this magnitude should be anticipated and prevented in a shift away from meat-heavy diets. Current estimates suggest food needs will grow by 70–100% by 2050 (from 2010 values) (Godfray *et al.*, 2010); such projections need to take into account food waste, optimal diets, food distribution, etc. In order to understand how this additional food would be produced and provided to those living at that time, let us take a quick look at the many varied dimensions of agriculture that must be orchestrated harmoniously to deliver increases in production and improvements in product without destroying the system (soil, water, biodiversity) on which it is based. Here is a such a list:

- Farmers/workers (wages, health and healthcare, retirement, security, and land tenure)
- Water Accessibility and Use
- Pest Management
- Weed Management
- Soil Quality (nutrient and biologically rich, toxin free, and stable)
- Seed Quality
- Biodiversity
- Sufficiency (Availability) and Accessibility
- Nutritional and Medicinal Richness
- Climatological Situatedness and Resilience
- Carbon Neutrality (or Reduction)

All of these components of farming demand serious attention. We need to make sure that the laborers are taken care of and treated humanely. It is definitely time to end the abusive practices that exist in current agricultural communities (low, unreliable wages, price instability and exposure to harmful levels of hazardous chemicals). Water must be collected and used in conscientious ways whereby aquifers are not depleted and water is not stolen from one community for the betterment of a richer, more powerful one (as is true in California/Mexico and Israel/Palestine today; see Hass, 2016). Pests and weeds must be managed without the addition of toxic chemicals but rather through companion planting, where one plant serves to deter pests to another, and diverse intercropping (which effectively confuses bugs). Soils need to be stable (or increasing) in terms of nutrient availability and maintained in a pro-microbial life environment (Biel, 2016). Seeds are the fundamental source of all future foods. As such, they need to be revered for their genetic diversity and associated built-in resilience rather than merely as a means to a highly-profitable end. Diversity needs to become an end in itself because of all of its inherent and extrinsic value — to reduce pest and weed stress, expand productivity of total food, and enhance aesthetic pleasure. Food needs to be grown in sufficient quantities to feed the entire human population, and accessibility to good, nutritious food must be demanded as a fundamental human right. The healthful qualities of food, both for caloric and nutrient consumption as well as medicinal value, should be considered. Foods must be grown with consideration and respect of a location's climate, while recognizing that pipeline climate change will make this a moving target. And, finally, methods of food production should rapidly transition to a negative GHG emission mode, requiring a significant shift from modern industrial practices.

These are obviously many aspects/dimensions that must be considered in the establishment of a "new," sustainable agriculture system. Yet, it would be wrong to address each of these individually as agriculture definitely needs a holistic approach as much as anything — some might cogently argue the specialization and compartmentalization of tasks/aspects of agriculture led, at least in part, to its current failings. Fortunately, much of the vision of how this might be done has been well-established, both in practice and in theory. For the past 10,000+ years, humans have been partaking in a grand experiment, and through this long process, successful models have been implemented. Connecting those methods with modern knowledge about health and ecology present us with compelling paths for future agriculture development — a truly sustainable and just 'green' revolution.

Agroecology provides a powerful alternative to current dominant agricultural practices. Some reading this will wonder where permaculture is within agroecology. It is a subset of it with clear design strategies; see Holmgren (2002) for an overview of the practice and Ferguson and Lovell (2014) for a current outlook. For our purposes we won't mention it explicitly but there is a great deal of overlap between permaculture and our form of agroecology. Agroecology looks at crop production within the context of 'environmental, social, economic, ethical and development issues' (Wezel *et al.*, 2009); FAO (2015b) provides a great overview of the principles of agroecology and its current activities worldwide. This broader lens forces analyses and practices that consider all the above dimensions critically and purposely. While agroecology continues to be an academic enterprise since it launched in 1930 (Wezel *et al.*, 2009), many efforts 'on the ground' represent applications of its theoretical and scientific findings. Both theory and practice have both come a long way to demonstrate the OWSP in the area of agriculture.

Agroecology doesn't conceive of agriculture as a poker game, where there are winners and losers (and in 'legal' gambling, *guaranteed winners* — those that own the casino and use identification 'cards' to manipulate people to extract maximum profits under the law). Agroecologists envision a society that values the incredible contribution that farmers provide and remunerates them accordingly. A relevant challenge is changing where food 'dollars' go. We spend inordinate amounts of money on food (see Chapter 3) and most of it doesn't go to the farmers. Many farmers are struggling financially, particularly small ones, and massive numbers are indebted because of the excessive loans they had to take out to get started or expand to remain competitive. The failure of our current system to provide farmers financial security (except the few that are producing a few subsidized crops) is a travesty. Your food dollars mostly don't go to them, whether you are eating a shrimp or drinking a cup of coffee (LaPorte, 2013). Organic methods of production now require more labor but since the labor is spread out over time (Pimentel *et al.*, 2005, p. 579) such methods can 'alleviate rural unemployment in many areas and … reduce the trend of shantytown construction surrounding many large cities of the developing world' (Badgley *et al.*, 2007, p. 94). Mann (2018) anticipates a higher commitment to agricultural labor in a future agriculture which functions guided by sustainable principles. Agroecologists emphasize the importance of purchasing local/regional food when it is in season — part of the bioregionalism agenda. This keeps more food dollars local where they can help support other components of local economies (Shuman, 2015), and discourages turning food into huge commodities for trade and profiteering, such as occurs on the Chicago Board of Trade[5.1] — a dominant exchange that demonstrates how powerful profit motivations are in determining what is grown and by what methods. Rather, in the agroecological perspective, food is understood to be the foundation of societal health and well-being, and as such, economic relationships among farmers, grocers, chefs, restauranteurs, and consumers foster living wages. Hence sustenance for all is encouraged.

Agroecological methods greatly reduce the need for 'outside' water (that which doesn't naturally fall on a field of crops). And given that

[5.1] CBT, which recently merged with another exchange to become the CME Group, Inc.

nearly 70% of current water use globally goes to agricultural activities (Clay, 2004), these reductions greatly enhance the availability of water for non-agricultural activities. These reductions also make communities more water resilient in the face of impending climate change. Agroecology promotes the capture of precipitable water through the building of holding ponds, channeling with tiles, collection by water barrels (connected to roofs by drains), and retention using well-managed soils. For instance, Cuba, a country that receives abundant precipitation (nearly all locations receiving 40 plus inches per year; Chamberlain, 1940), still needs to store water for times of drought — as of March 2017, many parts of Cuba were enduring one of their most intense droughts (53 out of 168 municipalities declaring extreme drought according to Martinez, 2017) — as well to enhance growing and production during seasons which tend to be water limiting. Other locations in the world with more water-limiting seasons or less precipitation than is typically associated with bountiful harvests of fruits and vegetables can improve production greatly through water retention (Kavdir *et al.*, 2014). Recently, it has been established that using desalinated ocean water can enable the growing of all types of food products even in deserts. For example, Sundrop Farms, located in a Southern Australian desert, utilizes indoor farming supplied by solar energy driven electricity for cooling and 'fresh' desalinated water derived from the ocean (Hrala, 2016; Stein, 2016). Aston University in Birmingham, England is researching the great potential for combining solar power and agriculture production (Aston, 2015). Connected to this project and cofounder of Seawater Greenhouse, Charlie Paton claims to have developed a technology that can deal with two 'seemingly intractable problems — a shortage of fresh water and brine discharge from desalination — thereby producing an elegant solution for crop cultivation, reforestation and realizing the value chain of salt, minerals and nutrients from seawater' (Paton, 2012). His Seawater Greenhouses in Abu Dhabi, Oman and Australia apparently demonstrate that it can be accomplished. Other technological solutions to 'low water availability' include the use of a water soluble, film-like membrane, developed by a Dubai-based company, Agricel, to grow crops with 80% less fertilizer and 90% less water than conventionally done (Parks, 2014). All of the above efforts also eliminate the need for pesticides as they are either exclusively indoors or have other means for keeping pests 'out' of their systems. These efforts show that growing food in some of the more inhospitable places can be achieved using new insights and technologies. Additionally, and importantly, expanding areas of food production into areas with low biological activity (i.e., low Net Primary Productivity) can facilitate the reduction in anthropocentric land use in other more biologically active regions thereby enhancing global biodiversity quite significantly.

Modern industrial farming has an addiction to fossil fuels and this must be cured. Food can be grown without the use of chemical killers (in all its forms — herbicides, fungicides, insecticides, rodenticides, etc.) or synthetic fertilizers. Humans have been doing so for more than 99% of the time since agriculture began. However, given the propaganda showered on the public, it is reasonable to understand why many people just do not believe it. Scientific research, however, demonstrates this, time and time again. Badgley *et al.* (2007) determined 'average yield ratios' (equal to 'organic' divided by 'non-organic' yield) in production in both developed and developing world contexts. Focusing on twelve categories of foods (including, grain products, legumes, meat and offal, and eggs), they found average yield ratios ranging from 0.816–1.060 (in developed countries) and from 1.57–4.00 (in developing countries). These valuable statistics led Badgley *et al.* (2007) to conclude that 'organic methods of food production can contribute substantially to feeding the current and future human population on current agricultural land base' (p. 94). Comparing organic managed corn and soybeans to conventionally raised crops, Pimentel *et al.* (2005) conclude that organically managed crops can produce yields equivalent to those conventionally managed, though they acknowledge that organically managed cash crops will have smaller yields over a longer time scale due to their 'dependence on cultural practices to supply nutrients and control pests' (p. 580). But Pimentel *et al.* (2005) also indicate that organic crop agriculture benefitted greatly from 30% lower fossil fuel inputs (for corn). Ponisio *et al.* (2014) conclude that a 19% yield gap exists between yields of conventional and organic methods of growing favoring conventional (using metadata with over 115 previous studies), a less sanguine result, but note that this gap can be reduced by more than half by incorporating standard agroecological practices of multi-cropping and crop rotations. Several researchers point out that recent funding has favored conventional methods and that if those precious resources were shifted to agroecological research, any yield gap would likely diminish further (Tilman *et al.*, 2002; Ponisio *et al.*, 2014; Biel, 2016). Additionally, factoring in the much larger yields of fruit and nut trees (measured in pounds) than conventional grains (such as wheat) on a per acre basis, Mann (2018) notes how potentially significant the shifting of crops might be in meeting the anticipated greater food demand.

There are many other reasons to think that yields of crops grown with organic, ecological methods will be more than sufficient for humanity's needs, including the adoption of unutilized indigenous farming technologies, as well as evidence that properly chosen strains of crops can produce favorably in the face of challenging conditions, e.g., Pascal Poot's tomatoes (Biel, 2016). For example, shade-grown coffee has indigenous roots and varies considerably from the coffee plantations of the industrial age. Figure 5.1

Figure 5.1. Coffee trees grown agroecologically in 'shade' under larger trees at Finca Coincidencia. (Photo: Carrington Morris)

shows an example of this on a Cuban farm. As Vandermeer and Perfecto's (2005) research demonstrates, shade-grown coffee farming has high biodiversity compared to monoculture coffee plantations.

Hence, one cannot stress enough how reductive our current perspective on food production has become. Yield per acre statistics drive almost all decisions right now and even prefigures many discussions regarding the viability of non-industrial forms of agriculture; this is in sharp contrast with many indigenous populations historically who had a very long-term focus on food production and, in some cases, maintained soil fertility for 1000 plus years (Biel, 2016, pp. 71–72). However, very sizeable and mainstream industrial organic efforts, such as that of Costco, now one of the top two sellers of organic food in the U.S., must also be interrogated (Fitzpatrick, 2015). As Solon (2015) points out, 'organic' doesn't mean that 'ecosystem multifunctionality' is maintained (as organic can still be done monoculturally and without much regard to the other organisms above and below ground). Nonetheless, we submit that holistic comparisons weigh very favorably on the side of agroecological methods even if yield comparisons are a bit more ambiguous. Muller *et al.* (2017) establish that organic agriculture is more than sufficient to feed future populations when decreased food wastage and reduced consumption of animal products are incorporated.

None of these pro-organic, pro-agroecological findings/insights should be surprising to anyone that has taken a historical look at agriculture. Yes of course there were famines in the 20th Century, and some were catastrophic, but these were driven by bad governance within undemocratic countries and profiteering — ironically, some of which were generated by the imperative to bring industrialization to the land rather than factors inherently tied to non-industrial methods of agriculture (Sen, 1992).

An agroecological perspective considers a farm as a producer of food but only so far as it is also a home for other species, a provider of ecological services, and a dynamic system continually reassessed and modified. Natural ecosystems have inherent strengths, such as 'interdependency, self-regulation, self-renewal, self-sufficiency, efficiency and diversity' (TWN and SOCLA, 2015). These are all properties of a perpetual agricultural system. Leary (2017) has provided a highly innovative approach that combines reforestation with agroecology. Agroecology reduces many harmful inputs, which does not only allow for production in a much healthier environment, but also creates conditions favorable for other life forms — thereby increasing total food available for humans and animals. Some have called agroecology 'regenerative', as well, as opposed to 'sustainable' or 'no harm'. In this mindset, one envisions a farm both as an entity that can produce food in perpetuity as well as a system that increases biological mass and biological capacity well above current levels found in industrialized systems. Estuaries, wetlands and tropical rainforests have been found to have the highest Net Primary Productivity (NPP) of all ecosystems on Earth (Miller, 1999); NPP represents the energy content of the carbohydrates produced by the plant via photosynthesis minus the energy lost during respiration. This maximum suggests that non-human ecosystems may have reached a biological 'healthy' limit. This level is several times higher than industrial agriculture (some 4 times higher according to Chapin III *et al.*, 2002, if above and below ground NPP is included) indicating that as an inventive species we are really only at the beginning of understanding what optimization really means and, particularly, how it may be achieved working with and within natural systems rather than against them (as synthetic pesticides, hormones, and artificial fertilizers clearly do).

Thus, these considerations point to the conclusion that food production could be significantly increased, using much less land than is the current practice, leaving a lot more for biodiversity to flourish (stewardship alone has its shortcomings), while at the same time doing it with greater vigor both for humans and all other species alike. To achieve this goal, we need to focus on keeping the soil rich in non-synthetic nutrients, full of microorganisms, along with earthworms, voles, moles, etc. Biel (2016) elaborates on the benefits of promoting the reduction of human labor in this regard, by his recognition of naturally occurring ecological pest control, an 'autonomous' system that kicks in when artificial pesticides are removed (p. 19). Farmers have for a long time understood the importance of using natural forms of fertilizer (such as manure and compost) to maintain a healthy, biologically active and physically productive system. For example, Japanese farmers have been successful raising ducks mutually with rice; the ducks eat weeds, weed seeds and the insects that harm the rice, while duck manure fertilizes the rice plants, and duck waste increases nutrient production on traditional rice paddy farms (Nierenberg, 2013).

Often land limitations are offered as a major obstacle to future increased food production. Agroecological considerations provide a multidimensional rebuttal to this as well. Mimicking nature, a central tenet of agroecology, enables the integration of crops, trees and animals. Trees such as the Acacia and Sesbania provide available fixed nitrogen to soil, which benefits nearby growing crops, stabilizes the soil and improves water retention (Nierenberg, 2013). Diversifying farms with plants growing at different heights and increasing the variety of foods being produced together, increases yields twice to thrice above the levels achieved when a canopy is not present (Nierenberg, 2013).

Growing indoors expands available 'land' and enables the expansion of the growing season to nearly year-round everywhere people live. Historically, soil has been the basis of nearly all forms of terrestrial agriculture — plants were grown in it and either fed to humans or fed to animals then to humans. Recently, driven by land access becoming increasingly limited (due to expansion in large agricultural farms) and high volatility in food prices, people have figured out how to grow many foods inside. Indoor growing takes many forms, sometimes with soil and sometimes without. High tunnels (greenhouse-like structures, of various sizes, with plastic covers rather than glass) increase temperatures year-round (up to 15°C above outside ambient levels) and allow for greater control of the environmental/weather conditions — by blocking heavy winds or preventing excessive rain to enter; on the other hand the increased temperature can be excessive, particularly in the warm months, but this can be abated somewhat by allowing outside air to circulate by rolling up the plastic sides. Loik *et al.* (2017) detail how Wavelength-Selective Photovoltaic Systems (WSPVs) can be used in concert with greenhouses to not only grow food (by allowing certain wavelengths through which are necessary for photosynthesis) but extract solar electricity for other uses. Pests can be better controlled as well, though not eliminated (grasshoppers and other insects find their way into the tunnels and can do considerable damage; for such issues, products like Jube, a biomimetic insect catcher that has shown promise in dealing with such critters, are possible solutions (watch the Biomimicry Institute's (2015) short video).

Figure 5.2. Growing Power (Milwaukee): The microgreen operation provides consistent income throghout the year. Greens grown include: sunflower, radishes, wheatgrass, beets, arugula, etc. (Photo: Peter Schwartzman)

Greenhouses can be used to grow highly nutritious microgreens (of sunflower, arugula, peas, radishes, mizuna, or any of another hundred possibilities) or just normal crops which benefit from the lengthened growing season (see Fig. 5.2). Hoop houses (just another word for high tunnels) can also be filled with aquaponic units (e.g., as found in Milwaukee's Growing Power farm) — three-dimensional systems that raise fish in large tanks while simultaneously raising herbs and vegetables in long flat trays above (see Fig. 5.3). These systems are 'closed loop' in the sense that the water in the tank (containing nutrient-rich fish waste) gets circulated to the plants above providing fertilizer to them, and then returns to the fish tank free of waste; more fully, the ammonium (NH_4^+) from the fish waste

Figure 5.3. Growing Power (Milwaukee): (top) Small aquaponics system with two growing levels above fish tank (above ground) (right) Large (60+ feet) aquaponics system with two growing levels above the inground fish tank below. (Photos: Peter Schwartzman)

is first converted to nitrite by beneficial bacteria, then to nitrate by other beneficial bacteria (these bacteria often thrive on the surfaces of small pebbles in the bottom of the tray) which then can be utilized by the plants. The crops, often something fast growing, such as arugula or watercress, absorb the nitrate and flourish. In this system, oxygen, usually pumped in via atmospheric gas, and fish food are the only external inputs. These systems have successfully raised many fish types, including tilapia, perch, koi, pacu, and bluegill, in tanks that are over 10,000 gallons in size — though smaller systems work can work as well. Fish varieties are chosen based on climate (some fish prefer it hot or cool) and market prices — tilapia was a very popular choice early on until their price fell abruptly as supply overtook demand. Indoor farming might also include chickens or shrimp, housed within a high tunnel or larger facility. Hydroponics — growing with nutrient-enriched circulating water rather than soil — is another means to produce large quantities of food indoors. Yet, Biel (2016) and others question these efforts that avoid soil, noting that soil contains thousands of organisms that have evolved symbiotic relationships that may be important nutritionally (for both the plants and the herbivores). For more on soil and agroecology see Engel-DiMauro's (2014) excellent treatment.

Vertical farms, though not exclusively indoor nor necessarily agroecological in scope, represent another huge untapped growth area in food production and another way to reduce land requirements. Both aquaponics and hydroponics can be expanded vertically, typically using artificial light, but vertical farming can take other forms and contribute to sustainable activities in many ways. Despommier's ground breaking book, *The Vertical Farm: Feeding the World in the 21st Century* (2010), makes clear how the future of farming might be viewed through a vertical lens. He projects that future farming will:

> 'include a building for growing food [plants]; offices for management; a separate control center for monitoring the overall running of the facility; a nursery for selecting and germinating seeds; a quality control laboratory to monitor food safety, document the nutritional status of each crop, and monitor for plant diseases; a building for the vertical farm workforce; an eco-education/tourist center for the general public; a green market; and … a restaurant. Aquaculture and poultry will be housed in adjacent but separate buildings with no physical connection to the vertical-farm building to ensure safety of the plants' (p. 179).

Furthermore, with appropriate design, these facilities can potentially not only capture sunlight for multi-level plant production, but will also capture it (and nearby wind and geothermal heat) for the production of electricity and heat, in climate zones requiring additional heat for its crops. Other research efforts have expanded on these potentialities. Touliatos *et al.* (2016) demonstrably found that the vertical farming of lettuce outperformed conventional hydroponics production. In Japan, indoor vertical farms of lettuce produce a remarkable 100 times more lettuce per day than conventional 2-D growing (with the same amount of land) while generating 80% less food waste (Macdonald, 2015). These findings are not surprising when one considers how vertical farming combined with ecological practices have the capacity to meet so many goals of sustainable agriculture. It enables the growing of more food on less land. It enables the three-dimensional farming of insects, though eating such animals are still a rather novel concept in many more developed nations despite having a long history in many other cultures (Menzel and D'Aluisio, 1998). Vertical farms can have some very interesting sustainable add-ons as well. Despommier (2010) points out their viability in: (1) filtering municipal waste water; (2) producing medicinal drugs (even as portable units during civil unrest or natural disaster) for humans (noting that Ayurveda recommends 315 herbal medicines as essential drugs); and, (3) a biofuel producer. With all these options, the future contribution of vertical farms remains bright and their full potential huge.

We should be clear that while vertical farms are a very intriguing development, they aren't without concerns as well. Some of the efforts thus far have required a great deal of infrastructure, which comes at a high cost, in dollars and materials. If external lighting is required or pumps and air moderation are also incorporated, energy expenses need to be accounted for. However, these can be potentially reduced with low cost solar electricity powered systems. Cox (2016) notes the limited variety of crops that might be grown this way and the early indications that any produce that is grown enters niche markets and, thus, does not meet the food needs of most people. Thus far, many of the farms that been successful have banked on higher price points to customers who 'are willing to pay a premium for foods they perceive as better for them' (Erbentraut, 2015).

Returning to the potential of insect biomass as a food source for humans, insect consumption has the potential to provide substantive nutritious addition to world diets and reduce land demands as well (Menzel and D'Aluisio, 1998; Van Huis *et al.*, 2013). Insects have just begun to make their way into some grocery stores in the U.S., typically as grasshopper meal or chips — for example, Chirp Chips (Chirp, 2017), which made a splash with an appearance on Shark Tank, and Cricket Flours, which has opened in Portland, Oregon (Crick, 2017). This is a particularly attractive step given the strong environmental advantages associated with eating insect protein rather than other meat forms. Crickets, for instance, require 1/9th of the plant input of cows (raised for meat) (de Nijs, 2016). Payne *et al.* (2016) have shown that crickets, palm weevil larvae and mealworm are more nutritious than beef or chicken, with a highly diverse composition in comparison with commonly consumed meats, suggesting that insect consumption may be 'effective in combating under-nutrition'. Importantly, as well, insect larvae happily eat farm waste as well as serve as high quality fertilizer, aspects that allow for circular processes, a key element in agroecology.

Fundamental to any sustainable food system are the seeds that serve as the next generation's resources. Historically, human diaspora owes much of its existence to seeds. Not only did they allow the introduction of known food sources but seeds themselves served as food that could withstand long voyages. However, more importantly, seeds permitted (and still do) the settlement of larger densities of humans into villages and shifted them away from a primarily nomadic existence. The future of humanity will likely rest on its ability to use seeds to generate food reliably. Yet, predicting which type of seeds will be of value generates a great deal of controversy. Some environmentalists passionately argue for the development and promulgation of genetically-modified (GM) seeds while others vehemently demand that we rely on more natural seed types, using selection methods that don't require directly altering the genetic makeup of these 'sleeping' giants, i.e., seeds that are dormant. As maintaining high levels of diversity represents a key goal of agroecology, it is seductive to think that complex genetic manipulations (something we have only been able to do for a very short period of time — gene splicing began in 1973) are necessary and justified. But as stated before, GMOs have been shown to reduce not only the genetic diversity within given crops but also across an entire ecosystem's biota, as they detrimentally impact pollinators and beneficial species as well as 'pests' (Scherr and McNeely, 2007). Beyond diversity considerations, the agroecological approach prioritizes methods that do not restrict ownership (of seeds or chemicals) to a few companies which limits options available to growers, thus disempowering them from doing as they think best. For centuries farmers have collected seeds from their farms for use in a subsequent growing season. They have chosen seeds that were robust yet ones that conformed to their needs and the market's wants — a mutualism, when working, that lifted both components of the system.

The seeds of tomorrow will need to be well suited to a region's soil type and climate. An agroecologically-informed system recognizes the value of holistic approaches, such as those that base the selection of seeds on a region's soil types and climatological conditions. And while modern agriculture often modifies (enriches) soils through artificial fertilization and increases natural supplies of water (e.g., by tapping into aquifers), this approach requires a lot of additional energy and other resources (such as irrigation structures). These modifications inherently create a system that is more dependent on outside resources, not less, once again a move in the wrong direction. Agroecology welcomes the selection of seeds that are well-adapted to existing conditions, using old traditional methods, as well as more modern (but not genetically-modified) approaches. However, we cannot resist the pun, climate change has and will continue to modify the 'playing field' that all crops grow in. Under these changing conditions, agroecology stresses that good communication channels are maintained between the scientific research centers and farmers as well as the consuming public, who will be eating or using whatever is grown (food, fiber for clothing, marijuana for pain relief, etc.). For example, the recent finding that climate change might induce plants to shift production towards more carbohydrates and less protein (consequently reducing nutritional quality) points out how new environmental conditions may have unexpected consequences (Ziska *et al.*, 2016). Incorporating agronomic research into the planning process facilitates the choice of better crops and polycultures, which do not require many additional inputs. Improving communication with the public about the food and nutrients they consume allows them to serve as more informed and participatory members of society.

Agroecology seeks to emphasize the value of perennials as they function more on natural inputs. With the waste streams of perennials in mind, The Land Institute (LI) started attempting to produce the world's first perennially grain more than 30 years ago. LI's founder, Dr. Wes Jackson, a geneticist, relocated to the Central Plains (Salina, Kansas to be exact), where he grew up, to begin this groundbreaking work. Years of diligent research led LI to patent Kernza®, a perennial wheatgrass that (along with other perennial grains in the works at LI) has the potential to displace modern GMO forms of grains and shift large scale agriculture away from its most destructive practices. There is potential to develop perennial grains of many sorts, including, but not limited to, oats, maize, barley, soybean, buckwheat, and wild senna (DeHaan *et al.*, 2007). One key benefit of perennial grains is their ability to utilize a greater proportion of sunlight, as current annual crops often don't have significant leaves for photosynthesis until halfway through a potential growing season (DeHaan *et al.*, 2007). This difference will undoubtedly allow for greater yields over longer time scales. Additionally, deep rooted perennials increase water access and retention (Glover, Cox and Reganold, 2007), as well as soil carbon storage. Concerns over perennial grains having invading impacts (in non-agricultural zones) are thought to be negligible given the characteristics of perennials (as compared to annuals) — i.e., noncompetitive, short plants, with large seeds (DeHaan *et al.*, 2007). Given that ~85% of the human 'food calories' currently come from annual crops (Land Institute, 2017), a shift to perennials could have profound positive implications. Importantly, Badgley *et al.* (2007) also find that production per area is greater on small farms in both the developed and developing countries supporting the polyculture emphasis of agroecology (p. 94).

Pre-industrial forms of agriculture had strong sustainable traits. For one, energy usage (total and per capita) was far smaller than current needs. For example, pre-industrial agriculture produced maize at 14–31 output/input energy ratios, while semi-industrial (animal assisted) had ratios of 5, and full industrial (fossil fuel and mechanization) has ratios of 2–3 (Leach, 1976). Of course, humans did most of the physical labor (and such a life of backbreaking work may not be ideal for most of us) but the past does suggest that current energy outlays for modern agriculture (almost all from fossil fuels) are excessive. Modern agriculture concentrates so much of its energy to produce grains. It has also converted almost all crops into annuals, despite most of the ancestor plants being perennials — this conversion started a very long time ago but it is almost complete now (Pollan, 2007). In the USA (2012), grains and beans (such as soy, corn, barley, rice, and wheat) and forage amount to 296 million acres, compared to 4.5 million acres used for vegetables and 5.2 million acres for orchard crops (USDA, 2014b). This means that more than 96% of U.S. crop land is used for annual grains. And of grains grown, it is estimated that >90% of this is either fed to animals or to humans as corn syrup (De Schutter, 2012). Additionally, nearly all grain (96%) fed to animals is corn (USDA, 2017d). Globally, nearly half of the World's cereal production is used to produce animal feed (De Schutter, 2012). Additionally, all of these grains are grown as annuals, meaning that huge inputs of energy and resources are required to replant each year's successive crop.

Once again, agroecology delivers solutions to the energy challenges raised by conventional farming. In an agroecological system, fossil fuel energy is reduced by reducing the distance food travels from farm to fork, as well as eliminating fossil fuel-generated nitrates, pesticides, etc. It promotes the eating of fresh rather than processed or 'artificially' preserved food (such as bananas that are harvested early and stimulated to ripen — off the tree — with questionable chemicals). It depends on natural fertilizers (from animal manure or compost) and biological pesticide agents (such as beneficial insects and polyculture methods). However, as shown in the last chapter, concerns over energy scarcity will be a moot point in a robust wind/solar energy transition. If sufficient solar-based energy is provided, nearly every agricultural need can be provided as well. There just is much more sunlight energy than humans could imagine using in the foreseeable future. And, as people in remote villages are finding, solar panels can do wonders to bring electricity and power to bear on rural needs (such as, lighting, cooking and heating), many of which have historically been very taxing on the local ecosystems because of their dependence on firewood.

In contrast to modern industrial agriculture, agroecology considers the nutritional and medicinal value a priority in what is grown. In this way, within an agroecological framework, it doesn't make sense to grow excessive amounts of corn for the production of high fructose corn syrup, in order to meet the created market for junk food (Schlosser, 2001). Sadly, our current industrial food system does just that. Actually, it does much worse. Through its senseless and selfish profit-seeking mandates, it creates an insatiable demand for its nutrient poor, disease inducing foods by spending millions of dollars each year advertising them, on TV and the Internet, on billboards, in newspapers and magazines, and as endcaps (products strategically positioned at the end of grocery store aisles, often established through contracts between corporation food purveyors and grocery store owners; Moorhead, 2015). Coupons are another way to cajole, especially economically-challenged consumers, to purchase highly processed foods and subsidized meats. High calorie foods (such as sodas, candy, cakes, and chips) are heavily marketed and accessible to all, even those living in poor urban areas where traditional grocers are less and less common and liquor stores and gasoline convenience stores often take their place. Low income folks receiving food stamps (or their equivalent) are often compelled to purchase these high energy, low nutrient density foods because of accessibility and the 'thirst' for calories (to ward off hunger pangs). Public schools, which receive federal subsidies, purchase specific foods, ones mass produced by industrial agriculture, often not the most nutritious and fresh. Given all of these forces working in tandem to push nutrient-poor foods, is it any wonder that we have such serious diet-related health problems? Is it coincidental that Americans have a lower life expectancy than most developed countries? Fortunately, the agroecological approach tackles each of these perverse incentives fairly simply. Diversity of crops increases the nutritional value of available foods. Vegetables in particular are the 'most sustainable and affordable way to alleviate micronutrient deficiencies among the poor' (Nierenberg, 2013). Integrating indigenous/ancestral foods into cooking classes and local restaurants has improved health and wellness among Native Americans (Nelson, 2013). Finally, thousands of plants have been shown to have favorable medicinal qualities. In fact, many active ingredients in pharmaceuticals currently are compounds found in plants. Greater recognition of the potential for expanding natural remedies will demand ever greater protection of remaining natural habitats, particularly in the tropics where endemism is common and diversity is greatest.

Similar to industrial agriculture, agroecology recognizes the importance of biological research. There is still much to learn about nutrition, medicine, crop rotation, etc., and ecology and molecular biology have a lot to offer us as we transition to sustainable agriculture. In fact, recent advances in genome sequencing provide a potentially powerful approach to identifying the best varieties to cross-breed (see Bevan *et al.* (2017), for an excellent up-to-date introduction to this topic). However, agroecology demands that such R&D not be done recklessly, as has been done with GMOs, and that it isn't done under corporate control where economic interests reign supreme and 'sharing' knowledge isn't welcome.

Furthermore, more research into the relative contribution of urban agriculture to food production and the specific agricultural activities engaged in needs addressing; as said by D. Schwartzman and P. Schwartzman (2017):

> 'focusing on smaller spatial scales, related questions include whether primarily urban/suburban communities can potentially feed their populations utilizing these food systems, and if this is the case what are the area/population ratios, and what are the relative roles of open field, greenhouse and vertical farming in meeting this need? Will harvesting ecosystems in rural areas play a significant role in meeting this objective? How porous should bioregions be with respect to food supplies? These questions...merit serious attention, moving from the abstract to concrete implementation.'

Any future agriculture needs to confront the reality of climate change and at least be carbon neutral. Biel (2016) demonstrates how sustainable food systems have the potential to enhance carbon sequestration from the atmosphere. This can be done by enhancing organic carbon storage in the soil with deep rooted plants (including many perennials as we have previously noted) and the use of silicate rockdust to speed up chemical weathering. Biel (2016) also suggests that grazing herds can be a climate mitigation strategy but we find that due to there being very little peer-reviewed research on this claim, the emissions of methane and nitrous oxide associated with such activity may negate its carbon absorption potential. We suggest instead that:

> 'any claim regarding net carbon sequestration from the apparent increase in grassland productivity, accompanied by the increase in cattle population should be demonstrated by systematic research that quantifies the flux of carbon into the soil and the increase in methane emissions to the atmosphere' (Schwartzman, 2017).

As also noted in Chapter 4, Hudak (2017) agrees with this position and expresses support for conversion of 'unproductive' cropland into forests, if such action can be shown to reduce atmospheric carbon over time.

Shifts in agricultural methods need be comprehensive. Moving some locations forward while leaving the rest as is will only reduce the problems, not eliminate them. Current global disasters created by industrial agriculture are not effectively solved farm by farm. Hypoxia in the Gulf of Mexico, driven by excess nitrogen runoff way upstream, represents a good example of this point. Excess nitrogen must be eliminated across the board, not just on a few farms. Tilman *et al.* (2002) note that a multitude of simultaneous technological methods can be introduced to increase fertilizer efficiency, however they also point out that natural fertilizers (from organic compost or green manures) release more slowly than conventional ones and this may also reduce the output of some crops. Recognizing the absolute need to reduce erosion (and hypoxia) points to the importance of integrating perennial grains on a large scale, as 'conversion to perennials was six times as effective at controlling erosion as was improvement management of annual crops' (DeHaan *et al.*, 2007). Other problems, including antibiotic resistance and the genetic drift of GMOs, are likewise not likely to be properly controlled by merely reducing their impact little by little.

5.4. Cuba, a case study

Cuba, which will serve as an inspiring case study of agroecology 'in the works' in Chapter 8, also provides a great example for how agroecology hypothetically could provide a nation a functional and just sustainable mode of agriculture. Cuba, an island nation (about the size of Virginia), has a population of 11.4 million people. Cuba, located mostly between 20–23°N as it is horizontally long and narrow, represents a subtropical nation and thus has only modest seasonal differences. Havana, Cuba's capital, located in its northwestern reaches, sees a high temperature range of 24–31°C, characteristic of a subtropical location. Its biggest seasonal difference is found in terms of precipitation variations. Its dry season (from November to April) gets 15 inches (37 cm) of precipitation while its wet season (May to October) gets 34 inches (87 cm) (Climatemp, 2017). These climatological conditions make Cuba ripe for growing all year round, though its mountainous areas and occasional hurricanes certainly challenge any farmer's success. And given its preferred latitude, the ample sunlight that it receives serves as an amazing resource for both crops and energy production (as we will see later).

Playing the role of a Cuban agricultural director (or council) for a minute, let us elucidate the decisions necessary to feed a country sustainably. First off, the director will want to ensure that there will be enough food for all of Cuba's people throughout the year. Typically, 'enough' incorporates supplying not only sufficient calories but sufficient amounts of protein as well. This goal might be reached by way of setting up stable supply and demand relationships — noting consumption rates of various food items and ensuring that sufficient acreage for each of these foods are in production, accounting, of course, for expected losses. She also must attempt to maximize the storage of food through preservation, as most regions of the World are not able to produce sufficient quantities of food throughout the entire year. Next, the director must consider how the country can do this in an economically affordable way. Initial logic might suggest that this would require Cuba to grow/produce all its food by its own enterprises (i.e., on the island and with Cuban labor). However, as we learned in our recent Agroecology Tour of Cuba (sponsored by Food First), this logic does not work as expected. Some products such as wheat cannot be grown effectively in Cuba due to climatological conditions and some crops can be grown so well in Cuba (namely, coffee) that its added value on the international market makes it economically sensible for Cuba to export its high-quality coffee while importing lower quality coffee for consumption by its citizens (The National Association of Small Farmers (ANAP), personal communication, June 2017). Next, and very importantly, the director would need to consider how these first two goals could be met without denigrating the environment and thus ensuring the capability of such economically-sound food production into perpetuity. Any agricultural system that doesn't meet this third goal is destined to fail, perhaps not in the near term, but definitely in the long horizon. Notice that the first three considerations are nothing more than what has come to be called the 'triple bottom line', where social, economic, and environmental considerations must all be fundamentally confronted to make the system truly work. However, there are other related concerns that the director must consider as the three important goals above are not sufficient in and of themselves. Among these, perhaps most important, is the establishment of a sustainable supply of farmers. Though technology has improved greatly, farming is still largely a labor-intensive activity (though robotic workers energized by the solar energy grid of the future will likely reduce laborious, monotonous tasks). And so long as it remains so, it is critical to maintain sufficient supplies of farmers and laborers, who not only know how to do the multiplicity of skills demanded on a farm but who also have the energy and motivation to do it; the latter of these, may be the greatest challenge as other profitable professions expand in non-agricultural areas. Relatedly, since agriculture includes more than merely growing crops and stewarding the land, one must also guarantee that the other components of the 'food to table' landscape (including, processors, chefs, distributors, etc.) are properly addressed as well. Additionally, the director needs to ensure that the best scientific knowledge and technological practices are properly communicated and taught to all those connected to agriculture.

With all this evidence establishing the beneficial properties of agroecology, most reasonably wonder why they aren't being employed. Others, recognizing the reckless profit motivations that drive our current unsustainable system, ask the next question, 'how can we move to an agroecological-based food system?' In Chapter 8, we will share some of the methods being used to move us in this direction as well as present examples of what modern forms of it look like, and indicate persistent key challenges. Knowing that it is possible is half the battle, implementation becomes the next great effort. (We note that Wezel's (2017) edited work, *Agroecological Practices for Sustainable Agriculture,* was not available at the time we wrote this chapter, but imagine it will have a lot of complementary elements.)

Chapter 6

False Solutions and Misdirections

6.1. Introduction

Given the profound insights and remarkable opportunities for transformation discussed in preceding chapters, one might reasonably wonder why we, intelligent primates, have not acted more quickly and decisively. Among the many plausible explanations for our complacency in the face of ecological duress exist a few systemic barriers and misdirections that require a careful analysis. Before we can move forward with grand initiatives to address such issues as climate change and industrial toxic waste, we have to come face-to-face with these great obstacles. With them clearly identified and deconstructed, we can better tackle the economic and moral challenges that lay ahead for all of humanity. Failure to do so will likely lead us down paths that will only delay progress and waste resources.

The following account does not pretend to be exhaustive in its consideration of those distractive elements alive and well in today's society. However, it does aim to identify some of the more critical ones. A movement capable of tackling the many seemingly insurmountable challenges faced by humanity will undoubtedly need to be united on several fronts. Huge resources (physical and human) will need to be put to the task of righting this ship. If we continue to go down paths that are heavily resource-intensive without clarifying the key obstacles or traps along the way, we will almost definitely waste time and limited resources. In a worst, yet conceivable, scenario, we might not have the time or resources to make more than a few mistakes without triggering a full-blown sixth global great mass extinction on Earth, something that esteemed biologist E.O. Wilson has claimed we were already in the midst of as early as 1992 (in *The Diversity of Life*). Honestly, we actually do not know where we stand in this regard but we can employ the precautionary principle and, paradoxically, simultaneously work from a position of great optimism while imagining the worst is yet to come. On this cynically ambitious note, let us begin.

There exist several critical challenges that we must fully deconstruct before we can make meaningful progress. Top among these obstacles is the modern Military Industrial Complex (aka the MIC). Incredible levels of wealth and capital globally are current siphoned off by this behemoth and, more disconcerting, there does not seem to be any end in sight. Second, slowing meaningful progress is the continued investment in many biofuels and nuclear power. Third, the effort to transition to renewable energy systems, while well-intentioned, still suffers from the privileging of one 'bottom line' (economic) at the expense of other ones (environmental and social).

In addition to these structural challenges, there are many ideological battles that we still face. First among these is the overemphasis on the importance of the size and growth rates of the global human population. The long-held tenet that considers human population size of preeminent significance, found in many environmentally-focused ideologies, definitely precludes a more meaningful (and reflexive) focus on those more impactful factors that contribute directly to the breakdown of ecosystems and human communities. In addition, ideologies that are inherently hateful, such as sexism, racism, xenophobia, and homophobia, all detract from a successful transition to a more harmonious world. Each of the issues warrants serious deconstruction, once realized can contribute to the redirection of huge resources and greater unity of progressive movements worldwide.

Obviously, there are many other obstacles to avert climate catastrophe and mass extinction, such as, legally treating corporations as people, 'free' trade agreements that trump national/local law and jurisdiction, hyperconsumerism, commodification of life, etc. However, many people now understand these obstacles and are in agreement that they must be deconstructed and overcome, despite their formidability. In sharp contrast, the obstacles we will spend most of our time discussing here are different. There are actual supporters of these obstacles among the 'rationally' minded and environmentally-conscientious. This realization makes it all the more important that we identify them as soon as possible.

6.2. The dinosaur in the room: the military industrial complex

In Chapter 4 we focused on the Military Industrial Complex (MIC) as a major obstacle to implementing a prevention program to avoid climate catastrophe, with an ever-shrinking window of opportunity. Here we look at its broader impacts on society. It is clear that any progress in our civilization's efforts to live at peace must deal with the extraordinary resources (physical and monetary) that militaries in the modern world consume. In a political environment where the lack of available resources (usually couched in monetary terms) is used to justify austerity measures in nearly every aspect and sector of our society, it is quite amazing that military budgets, particularly in the most powerful nations, continue to rise. More phenomenally, this rise gets almost no critical examination within mainstream political circles or, shockingly, within most environmental discourses. Increases in military budgets in the era of ubiquitous 'terror' appear to be considered inevitable. Such ridiculous circumstances demand that we first identify how much money is used to 'safeguard' the world. Next, we must reconcile the energy demands of these seemingly resource heavy components of our civilization.

To get a grasp of how big a drain exists in the MIC's consumption of public resources, one need not look any further than the Stockholm International Peace Research Institute (SIPRI) which has been studying the matter of military expenditures since 1967 (www.sipri.org). SIPRI, considered the most authoritative voice in this area, has kept very consistent data since 1988 for approximately 171 countries around the World. Each year they publish a Yearbook which gives an overview of the latest figures and trends. In 2014, world military expenditures were US$1.75 trillion — that's US$1,750,000,000,000 (for those that like to see big numbers written out), or 2.3% of the World's gross domestic product

(GDP)! There were 7.3 billion humans alive on planet Earth in 2014 (World Population History, 2018). Thus, globally, per capita military spending was ~US$243 in 2014. And although globally military spending has been declining a little over the past three years (–0.4% between 2013–2014), some countries (most notably, Saudi Arabia and Russia) have made significant increases recently (Perlo-Freeman *et al.*, 2015) and the U.S., under the Trump Administration, has recommitted to even more dollars for the military. Also, sadly, despite all the pretenses of 'peace', global military expenditures have actually gone up 17% from 1988-2015 (in constant 2015 dollars) (Perlo-Freeman et al., 2015). The US$1.75 trillion being outlaid annually may sound like a very large sum (if you look at the sheer amount) or a relatively small amount (if one focuses its percentage of global GDP), but, either way, it is a number that must be understood for what it does, and what it prevents, directly or indirectly, from happening as well. Big numbers are hard to work with and most of us do not live in the world of millions (and certainly not in the land of billions or trillions). In fact, this realization alone may explain why so many feel so detached from the budgets of large governments or militaries. So how much money is this really? In 2014, U.S. military spending was US$610 billion. This is nearly six times larger than its education budget (US$102B) and seven times larger than its transportation budget (US$85B) (Gould and Bender, 2015). But perhaps the most outlandish statistic is this: the U.S. military budget is larger than that of the next six largest national military budgets combined. In 2014, China's allotment was US$216B, Russia's was US$85B, Saudi Arabia's was US$81B, France's was US$62B, the United Kingdom's was US$61B, and India's was US$50B; these six nations' total stood at US$554B, and the top seven spent US$1.164 trillion or 65% of the global total (Gould and Bender, 2015). Also, the U.S.'s military budget has grown an astounding 108% since 1988! So much for the end of the Cold War! Clearly, it isn't just that there is a lot of money spent on the military, but that the distribution of such spending is greatly imbalanced; a few nations dominate the spending and use it to further their own interests to the detriment of most humans and the planet. Consider how much foreign monetary support from the U.S. is actually in the form of security aid and direct arms sales each year. Based on known data, nearly US$10B is spent in security aid (requested in 2017) and US$22B in arms sales (in 2015) (Bearak and Gamio, 2016).

Despite the outlandish sums spent on the military in the United States, the conversation over limited economic resources nationally and locally almost never touches on this dominant drain in our society. In the early 1980s, when the story of the extraordinary high prices for military parts broke (see Barron's piece in the *New York Times* of September 1983), many Americans cried 'wolf'; back then even the iconic news program *60 Minutes* did groundbreaking journalism on the topic. Thirty plus years later and spending on U.S. defense on the order of several hundred billion dollars a year continues unabated, and yet one hardly hears so much as a peep from anyone. It is almost a conversation non-starter given all the emphasis on the news of mounting terrorism and global instability, and the perpetual 'War on Terror' commenced during G.W. Bush's Administration. To date, the U.S. has spent an estimated $4.3 trillion on wars in Asia and the Middle East (Crawford, 2017). Recent conversations we have had with highbrow progressives leaves us to conclude that the modern incarnation of the MIC is perceived as just too powerful to be dismantled. Even Naomi Klein, while pointing to the financial impact of a 25% reduction in the top ten military budgets (which would free up US$325 billion dollars), describes it as, 'probably the toughest sale of all, particularly in the U.S.' (2014). This conclusion (significantly premature in our minds) leads many to ignore the great 'straw sucking' of resources that the maintenance of such a behemoth requires. However, perhaps tides are changing, Bill McKibben, founder of 350.org, recently wrote in an article entitled, 'A World at War':

> 'Carbon and methane now represent the deadliest enemy of all time, the first force fully capable of harrying, scattering, and impoverishing our entire civilization… [Global warming] *is* a world war. And we are losing' (McKibben, 2016).

Like Malm (2017), McKibben goes on to talk about how we need to mobilize like we did in World War II — in this case to confront the threat of C3, instead of the Nazis. And while he does not mention actual dollars that could be redirected from the MIC, his willingness to hint that there might be a relationship between the MIC investment and a reinvestment in renewables is a positive sign. Yet, the connection needs to be made more explicit and it needs to be a key point if we are going to make headway. Why it took this long for it even to be hinted at suggests the myopia of the mainstream environmental movement, one that must be overcome.

Beyond the sheer resources consumed by the military, the dominant military paradigms proliferated around the world today must also be challenged in order for us to move towards peace. Current allotments for 'peace' are currently woefully insufficient and ineffective. Consider that the United Nations' 2015–2016 budget (for some reason they report it on a fiscal year rather than a calendar year) was US$8.3 billion. Notice, how this seemingly large number is a pittance compared to the World's 2014 military budget, cashing in at less than a half-percent of the larger total (that is less than 50 cents on every US$100)! Given all the rhetoric about 'peace', we are clearly not acting 'peaceful' in anything coming close to a collaborative, holistic way. Individual countries are exhausting their limited resources to promote policies that, in theory, protect their individual interests above all others. If history and recent environmental realizations, such as the 'boomerang' effect, teach us anything, we must work together to find solutions that consider all of us, not ones that stigmatize 'others', marginalize other 'others', or attempt to destroy others (using drones, cluster bombs or white phosphorus), all which fuel violent terrorist responses in the U.S. and abroad.

Not only are pro-military worldviews justifying the continued use of brutal and ever more immoral practices, but they have been used as powerful metaphors to tackle many of the other problems humanity faces. We have been addressing hunger for the past 70 years or so with a 'kill everything' monoculturally-driven, genetically-modified, agricultural mentality, one that largely ignores the synergisms and mutualistic relationships between organisms, and treats complex soil as no more than a mixture of elements (see extensive discussion of this mentality in Shiva's (2008) *Soil Not Oil* and Steingraber's (2010) *Living Downstream*. Industrial agriculture's insatiable need for the use of fossil fuel-based farm inputs (particularly, fertilizer and pesticides) draws directly from the production infrastructure of highly toxic chemicals that began development during

World War I and became refined and highly expanded during World War II (Steingraber, 2010). Post-World War II, with all of the militaristic machinery and chemical factories created during the war, saw the proliferation of derivatives of these chemicals and petroleum-powered machines into our agricultural spaces, enhanced during the two 'Green Revolutions' that were considered by many environmentalists at the time as a panacea for the billions of hungry mouths that had been born since the 1940s. The Earth is literally coated with these chemicals now because of this simplistic, militaristic approach to food growing, often promoted by 'peace-loving' people. The medical costs of this contamination are huge and the horror stories are endless.

The U.S. prison industrial system (some would say 'complex', so we call it the PIC) also shares much of its agenda with the MIC. Bigger is once again thought to be better. 'Larger prisons keep us safe' and more powerful armaments do the same. Being 'hard' (read, 'brutal') on criminals stems from a false belief that hurting someone (enacting revenge) is the best way to 'fix' them, a belief similar to one strongly espoused by military leaders today; do you not recall the 'Shock and Awe' of the beginning of the Iraq War (read, 'Invasion'). Is it any wonder that the country that seems to declare itself the 'freest' (namely, the U.S.) is also the one that has the largest prison population per capita by far? The U.S. imprisons 666 out of every 100,000 of its citizens, an incredible ratio compared to other 'advanced' or large countries; Japan's value is 45, China's is 118, India's is 33, United Kingdom's is 146 (ICPR, 2016). Clearly, a country that locks up (and often throws away the key for) so many of its citizens, disproportionally those of color, cannot claim to be the 'freest' or the 'most democratic' or the 'most peaceful'. Other countries also treat their prisoners brutally and most fail to consider criminals as mere byproducts of an unhealthy system of failing institutions. As a result, huge human resources are wasted in the PIC, in the U.S. and elsewhere, and the associated loss of human potential in greatly underestimated.

Ultimately, all this spending on the military and prisons could be greatly reduced if done with a different, non-militaristic, ecological approach. Rather than provide lots of details for this now (this is developed in other chapters), the point needed here is to emphasize how all-encompassing and normalized anti-peace and anti-environment (and thus 'anti-life') mentalities have become, in part, due to the insatiable memes of war and domination, less we not forget the hyper-masculinity associated with these dominant paradigms.

The redirection of military monies to other ends could have huge implications. Even if a small portion of the World's military budget were invested in sustainable infrastructure, we're talking about an extra couple of hundred billion dollars a year, it would make a substantial impact. Jeffrey Sachs, a world-renowned economist, back in 2005 computed that US$160 billion a year would 'cover basic needs [of the World's poor] in health, education, water, sanitation, food production, roads and other key areas' (Sachs, 2005). Therefore, with what amounts to a 9% redirection in global military budgets, we would have the resources to build the foundations for a global civilization at peace. What a bargain!

In closing on the matter of the MIC, and its PIC fallout, it is frightfully clear that there are enormous resources being squandered right now and 'staying the course' in the 'business-as-usual' models will continue to move us farther and farther from global peace and harmony. Arguments that couch their resistance to moving forward with sustainability-related initiatives based on the lack of sufficient financial capital without pointing out this grand misallocation of resources are specious and should be identified as such. **The continued unwillingness of 'progressive' voices to question the outrageous largess of world militaries prevents us from making the necessary redistributions of money and time.** It is important to add that these negative arguments happen at national as well as local levels and, as such, the 'calling out' of them needs to happen at all scales of society. Locally, people have the opportunity to see and understand better where their community's limited resources are being spent. Failure to identify that national and state contributions to these revenues continue to be cut while military spending continues to rise allows the status-quo to continue. Without these reorientations of our collective priorities and investments, we will be trying to fix our problems working with much less than a full deck. This is certainly not something an intelligent species, who wishes to survive well into the future, should be doing.

Now, assuming we make considerable headway with the MIC conundrum, might we still be stuck in metaphorical quicksand when it comes to fueling our expanding global population, which continues to consume more and more resources? Notice that this simple question identifies three additional challenges to humanity — energy, 'overpopulation', and physical resources — all notoriously mentioned as unsurmountable by some. What do we know about the future of humanity? If it survives, it is reasonable to assume that it will consume significant amounts of energy. Yes, there are some, such as Derrick Jensen, who argue that modern civilization is inherently unsustainable and imagine a future for humanity that rejects modern society, with its energy intensive elements and large cities, or others, such as Garrett Hardin, who believe we must accept a 'dog-eat-dog' world where all are always struggling 'to have' what they need on a finite planet. Unfortunately, they often make exaggerated claims about the limits regarding the planet and its systems, including human ones. These pessimistic views serve as powerful metaphors and therefore have considerable support among activists and radicals. Yet, as we show time and time again throughout this book, great opportunities exist, undermining their pessimistic arguments. Here we intend to establish, once and for all, that: (a) some energy directions propounded by activists and environmentalists are inherently unsafe, and therefore unnecessary in the transition; (b) the planet is not overpopulated with humans and will not be, perhaps ever; (c) many necessary resources, while limited, can be harvested in a sustainable way through recycling, reuse, and closed cycle processes.

6.3. Nuclear power redux

Throughout modern history, almost all communities have moved in the direction of greater levels of energy consumption *per capita*. And with the human population continuing to grow well into the future and with affluence growing even faster, the 21st Century definitely looks like one destined to continue this trend. If we are going to consume more energy as a population, we will have to get it from somewhere. As we have demonstrated earlier (Chapter 4), we can provide all the energy humanity will ever need in perpetuity using only solar based forms. Yet, while this is widely recognized (Zweibel *et al.*, 2008; Jacobson and Delucchi, 2009; P. Schwartzman and D. Schwartzman, 2011), there are still many prominent scholars (such as, former head of NASA's Goddard Institute for Space Studies, James Hansen) who argue quite vehemently that a transition to a fully solar sustainable energy system will require the use of nuclear energy and/or biofuels. Obviously, we think otherwise, but the emphasis of this misdirected view, especially in the 'mainstream' movement, does significant damage by diverting attention and unnecessarily undermining support for a transition to 100% solar-based renewable energy (RE). And while The Solutions Project (thesolutionsproject.org) and others understand this, too many do not and so it deserves a full hearing here.

Emphasizing the necessity of more nuclear power to 'fuel' the transition ignores the feasibility of transitioning without it and further locks us in to a highly secretive and centralized mode of energy production and distribution. As we have argued in Chapter 4, no additional supplies of nuclear energy are needed at any point in the transition to 100% solar-based RE. So, if it is not needed, why use it? There are generally three pro-nuclear arguments provided. One, recent technological advances in nuclear energy make them fool-proof and extremely safe. Two, the most influential argument on current debates on this topic is that nuclear energy is the best transitional energy source from a climate change perspective. And, three, we have so many nuclear plants now (especially, in France and the U.S.), we might as well use their electricity production as long as we can. While seductive, we find all of these arguments flatly misinformed. In chapters 3 and 4, arguments two and three are addressed.

Are modern forms of nuclear electricity production inherently safe? Our response is that this claim is highly problematic. The arguments for 'absolutely safety' surrounding nuclear power go back to its origins. As we have seen over the past 60 years, particularly with Fukushima, Japan (2011), Goiania, Brazil (1987), Chernobyl, USSR (1986), Three Mile Island, U.S. (1979), Kyshtym, USSR (1957), and Sellafield, U.K. (1957), nuclear accidents do happen and when they do, the results are far from trivial. In addition, huge (and rising) amounts of nuclear fuel/waste, which come by way of normal operations of nuclear power plants, have no permanent, safe home; case in point, the multi-decade debate and inability to utilize Yucca Mountain (Nevada) for all U.S. nuclear waste. We gain insight about how secure the nuclear power industry truly thinks it is by looking at its political history. The creation of the Price-Anderson Nuclear Industries Indemnity Act of 1957 (commonly referred to as just the Price-Anderson Act and renewed many times since) serves as the 'smoking gun' of the true risk understood by the industry. This Act greatly limits the liability for any accident at an operating nuclear power facility, and displaces the bulk of the financial risk to the U.S. government. There is not a better piece of evidence that communicates the inherent risk involved in using nuclear technology/fuel in the future.

And though all the above should be sufficient to move quickly away from nuclear energy, perhaps the biggest reason for doing so is related to the fact that the nuclear industry is integrated into the MIC, and the threat of nuclear attack is a long-standing instrument of imperial policy, as introduced in Chapter 4. Not only are nuclear missiles expensive (US$2–100 million apiece; Gronlund, 2013), they inherently put all humans at great risk of catastrophe. As we previously mentioned, the possibility of a nuclear war continues, with nuclear weapons now being possessed by nine nations, including notable hotspots such as the Middle East, with Israel possessing several hundred deliverable nuclear warheads. Nuclear power production necessitates the creation of nuclear fuel components that have inherent ties to the production of nuclear weapons. This nuclear fuel, which is used in existing nuclear reactors (which produce electricity), and the waste products of these reactors (namely, plutonium) can be reprocessed and formed into material that can be used to make nuclear weapons. Shocking to some, only a few kilograms of these inherent byproducts of nuclear-produced power are necessary to make nuclear bombs of tremendous power. *Countdown to Zero* (a documentary film released in 2010) tells of the rich history surrounding how easily such small quantities of such nuclear 'waste' have been repeatedly stolen (even by workers at nuclear power facilities, a few grams at a time) and sold secretly to non-governmental organizations. Additionally, by interviewing many of the political leaders and heads of the intelligence community during the Cold War (including, President Carter, President Gorbachev, and Robert McNamara), *Countdown to Zero* tells of the many times that a nuclear exchange between nuclear powers came within minutes of being called. Clearly, our civilization is not mature enough to handle the inherently secretive and innately hazardous world of nuclear power generation; enter President Trump into the equation for a 'slam dunk' demonstration of this point.

Notwithstanding all of the above criticisms of nuclear (and additionally the rejection that it is climate 'neutrality', as shown in Chapter 4), last but not least comes the simple point: the only nuclear energy we humans need is at a safe distance, 93,000,000 million miles away — our Sun with its fusion reactor at its core.

6.4. Biofuels?

Another huge distraction from an efficient and swift transition to renewable energy comes from biofuels. As some have suggested, the story should go as follows: particular forms of the world's biota have incredible potential to displace petrochemical liquid and gaseous sources of energy. And furthermore, the biological waste streams that we have created can also contribute mightily in this effort as well. And based on these two profound insights, we should redirect incredible levels of resources from non-renewables to these non-fossilized organic biofuels. Unfortunately, this story's falsity has failed to stop it from happening over the past ~20 years. In the name of 'greening' the Earth in this way, we are destroying it.

There tends to be truth to every story and this one has some truth. It is true that biological crops have a great deal of energy in them, something that historically humans have learned to consume, to stay warm and cook food. It is also true that many plants, serving as feedstocks, can be converted into liquids and gases that can be used as fuel. These biofuels rather easily substitute for petrochemical liquids and gases that run modern industrialized machinery and transportation systems; we knew about ethanol as a biofuel long before tetraethyl lead was created (and patented) as the 'must' anti-knock additive to gasoline (Black, 2008). It is also true that the dominant economic systems and the organizations (like the WTO) and corporations (like Archer Daniels Midland (ADM)) that operate within them dictate that commodities go to the 'highest bidder' and, thus, food can become fuel at the signing of a 'free-trade' agreement or a law providing government subsidies. In these cases, economic transitions proceed with virtually no concern for who might go hungry or how much land gets damaged. These facts do not mandate the expanded use of biofuels, however.

Fiction comes with falsities and delusions and the biofuel story is full of them too; see Holt-Gimenez (2007) for a quick, yet forceful, introduction. Firstly, trading land that provides food for land that produces fuel (for cars, planes, etc.) is plainly immoral, since it undoubtedly increases food insecurity. We need solutions that help the hungry get more stable supplies of nutritious food, not ones that exacerbate their hunger. Secondly, land use for food consumption should be better managed to promote biodiversity, GHG sequestration, and soil stabilization. Sadly, the 'hunger' of profiteering capitalists who seek biofuel revenues (often under 'green' pretenses) has transformed biologically-rich parts of the globe (particularly, tropical rainforests) into agricultural plots for growing sugar cane and other crops for ethanol production. Thirdly, biofuel production from industrialized farming only increases GHG concentrations in the atmosphere. While biofuel use generally does reduce GHG emissions when compared to fossil fuel use, these reductions have often been exaggerated. Liska *et al.* (2014) found that the production of corn residue which dominates cellulosic ethanol biorefineries in the U.S. results in the loss of organic carbon in the soil and increases atmospheric CO_2 levels. Charles *et al.* (2013)'s International Institute for Sustainability Development's extensive 2013 report on biofuels offers a similar conclusion regarding carbon dioxide 'savings'. In particular, Charles *et al.* (2013) find that the extensive biofuel subsidy programs offered in Europe actually cost motorists much more in fuel costs, and that governmental subsidies outpaced industries investments quite substantially. This study demonstrates that biofuels, and ethanol in particular, have been promoted as 'climate friendly' worldwide without sufficient analysis. And while certain biofuels, such as biodiesel, cellulosic ethanol, and algae-based fuels, have shown the ability to significantly reduce GHG emissions, we note that they are still producing GHGs when used. In contrast, solar-based energy forms are typically much lower emitters of such gases, as shown in full life-cycle analyses (Schlömer *et al.*, 2014). Fourthly, the efforts made to boost biofuel divert investment that could be better spent elsewhere. On this point, it is instructive to witness how quickly and with what firepower multinational agriculture firms, governmental agencies, and transnational economic interests were mobilized to make biofuels such a huge profit stream. The U.S. Energy Independence and Security Act (EISA) 'mandates production capacity for cellulosic ethanol and advanced biofuels to be 61 billion liters per year (bly) and 19 bly, respectively' (Liska *et al.*, 2014). How is that for a boost? Clearly, when corporations and government collaborate (collusion, in the case of ethanol, may be a better word) stuff happens.

However, some biofuels may be recommended. For example, capturing grease from fast-food restaurants for the purpose of reusing it to propel a lawn mower makes sense. Collecting other biomaterial waste, say lawn clippings or food waste, for the purpose of providing nitrogen rich sources for agricultural systems makes a lot of sense, but this is not an example of biofuel use, rather biomass recycling.

In the case of biofuels, unfortunately, the forces promoting them so far have prevailed and we have lived through a 20 plus year period where they have been highly subsidized; ethanol production tripled globally (primarily in the U.S. and Brazil) from 2001 to 2007 (Tenenbaum, 2008). From 1980–2009, in the U.S., biofuels received more than five times more subsidies than wind and solar combined (Johnson, 2011; Parkinson, 2016). These subsidies arguably only slowed the transition to wind and solar power, caused price fluctuations in food markets (as corn went from feeding pigs to 'feeding' car engines) and hunger in developing countries (Tenenbaum, 2008). Despite these findings, we still find biofuels promoted as part of the future energy portfolio for the World. And though, as we have argued, limited biofuel usage makes sense, the necessary distinctions are not being made explicit for how this limited use is starkly different in scale compared to the colossal expansion of biofuel use that has been thrust upon the World in recent years. And given the unrelenting efforts of multinationals to obfuscate the realities of biofuels (a simple Internet search on 'biofuel' myths will result in dozens of treatises discounting all the arguments presented here), we need to be particularly unambiguous and demonstrably clear about the role that biofuels should play in the coming years, as well as recognize their low efficiency of solar energy capture. Even the most benign sources of biofuel energy do not have the capacity to supply most of the energy required for our 21st Century civilization, in contrast to a wind/solar power infrastructure. While one could speculate about the future creation of high-efficiency capture of sunlight by genetically-modified photosynthesizers, high-efficiency wind/solar technologies are already available and being rapidly implemented for current society.

To be fair, we must similarly scrutinize wind and solar power implementation. We begin by pointing out that serious problems can even arise when wind and solar energy installations are not done with sustainability's triple-bottom-line at work. Over the past ten years, improvements in wind and solar technologies have made them economically-competitive, or even economically-advantageous compared to fossil fuels. While exciting, this economic reality needs a critical analysis. For example, hundreds of thousands of wind turbines are being made and being installed throughout the world, but are they being made by workers that are paid a living wage? Are the base compounds and materials, such as neodymium, used to make the components found within the nacelles of these turbines, being mined with adequate protection to workers and the environment? Given that wind turbines are being produced by many large multinationals (Siemens and General Electric among the top three producers), are humanitarian and environmental concerns as well as fair labor practices priority concerns in the production and installation of these turbines? To the extent that environmentally-minded individuals/groups have long privileged economic arguments over ethical and environmental ones in order to advance their grand proposals for renewable energy systems, they have fallen into and capitulated to adopting the same myopic economic principles that nearly always push holistic concerns to the margins (in the form of externalities). For example, many environmentalists currently boast about the fact that wind-generated electricity is now competitive with coal-based electricity on cost alone. And it is precisely this economic reality that is the major driving force in the very fast rates of installation of wind (and increasingly, solar) energy capacity. Yet, once again, if this economic battle is being won while still contaminating surrounding communities with toxic wastes from industrial mining, we should not call this transition sustainable in critical aspects. As we emphasized in Chapter 4, a strong environmental/ecological/health protection regime for renewable energy production and implementation must be coupled with its massive global expansion as well as the rapid shut-down of the fossil fuel energy supplies.

Where does this leave us? Must we only accept 100% morally-unobjectionable solutions (assuming they could be imagined)? Must we always compromise? And if so, where do we draw the line? We have addressed these questions elsewhere in the book (Chapter 4), for now, let it be sufficient to remind us how significant this dilemma is.

6.5. Is the World Overpopulated?

Nearly fifty years ago, Paul Ehrlich published, *The Population Bomb*, a widely popular (purportedly) non-fiction text that captured the imagination of many Western thinkers. That same year, 1968, *Science,* the most prestigious scientific journal in the United States, printed Garrett Hardin's highly influential article, 'The Tragedy of the Commons' (see Hardin, 1968). And over the next ten or so years, many other neo-Malthusians joined the fray. The Club of Rome's highly publicized 1972 report, 'The Limits to Growth', concluded that:

> 'Demographic pressure in the world has already attained such a high level, and is moreover so unequally distributed, that this alone must compel mankind to seek a state of equilibrium on our planet' (Meadows *et al.*, 1972).

As the 20th Century entered its last quarter, population-reductionism reached a fevered pitch, dominated by a belief that all the World's problems could be solved if we only focused on the true cause of environmental degradation, the 'swelling' populations of the 'Third World'.

Focusing on the 'Third World' was very convenient for those in the global North, particularly white males at elite institutions of 'higher' learning and wealthy capitalists. Rather than focus attention on rapacious industrialization, or the legacy of colonization, or the impact of imperialism on the global South, the sheer size and growth of human populations were deemed the drivers of continued human suffering and ecological destruction. Similar to Thomas Malthus, who almost 200 years earlier laid the groundwork for this ideologically driven perspective, with his *Population Essays* first published in 1799, many key Western environmental leaders stressed the 'natural' relationship of population expansion and resource depletion. Malthus, who has been largely misinterpreted by populationists, apparently did not underestimate the human ingenuity to barely meet increased needs of an increased population. He was more interested in providing a polemic that argued hunger and poverty were inevitable and, therefore, 'poor laws' (the 'welfare' of the time) were not effective measures to use in England or abroad. To sum up their argument in the modern discourse: humans are inherently driven by biological forces to reproduce beyond their means, ultimately surpassing the carrying capacity of the planet and causing mass starvation on a scale never seen before.

Sadly, this paradigm permeates much of today's environmental literature. Jared Diamond's *Collapse* (2005) (who also authored, *Guns, Germs & Steel* (1997)) and Derrick Jensen's *Endgame* (2006) are two great examples of how the neo-Malthusian perspective lives on uncritically. Jensen's 'Myth of Growth' clarifies where this perspective leads (i.e., to dismal ends):

> 'There are simply too many people. You've seen the pictures. Crowded streets in Calcutta, impoverished babies with huge hungry eyes and bloated bellies in Mexico, refugee camps in Africa, masses of Chinese crammed into filthy cities. The earth can't support these numbers. Something's got to give.
>
> And you've heard the arguments. The United States needs to close its borders to immigration from poor countries. Having finally gotten our own birthrate down sufficiently to more or less stabilize our population, the last thing we need is a bunch of poor (brown) people moving in to crowd us out (we know, also, that once they're here they'll breed faster than we do, and soon enough will outnumber us).
>
> I often respond to this argument by saying I'm all for closing the border to Mexico (and everywhere else, for that matter, all the way down to closing bioregional borders), so long as we close it not only to people but to resources as well. No bananas from Mexico. No coffee. No oil. No tomatoes in January. Many of the people who leave their families in Mexico (or any other impoverished nation) to come to the United States to work do so not because they hate their husbands or wives yet have not gotten to the point in their therapy where they feel comfortable expressing (much less acting on) this.... They come, one way or another, because the integrity of their resource base and their community (insofar as there can

meaningfully be said to be a difference) have already been compromised: the resources have been stolen, and the community is unraveling' (Jensen, 2006).[6.1]

To be sure, there exist many works that successfully critique the shallow nature of the neo-Malthusian perspective: Barry Commoner's *The Closing Circle* (1971) and *Making Peace With the Planet* (1990), Amartya Sen's *Development as Freedom* (1999), and, more recently, Angus and Butler's *Too Many People?* (2011), come to mind. These works point out how population growth rates decline in response to more equitable and higher living standards, as in 19th Century Europe, improvements in medical care, more occupational opportunities for women and better access to contraceptives throughout the world, as observed in the second half of the 20th Century. But despite the cogency of the arguments contained in these deeper, more holistic books, they are not what dominates the perspective of most Americans. And, the mantra of 'population' being the end-all-be-all driver of bad outcomes continues to be renewed and revived (consider Alan Weisman's *Countdown* (2013) and Stephen Emmott's *10 Billion* (2013)). Unfortunately, these largely one-dimensional books do significant damage to the ability of articulating a clearer vision of what is actually driving ecological collapse and human suffering, and, in so doing, makes it challenging to address the key causal agents to both human and ecological suffering, namely economic and social inequality, militarism and overconsumption.

Yet, above and beyond the monumental distraction that these population-reductionist works provide, they actually promote and reify other grandiose and often sexist perspectives. Almost all of the early works of the population gurus (of the 60s, 70s and 80s) put almost all of the attention, and, therefore, the onus, on women. Huge efforts (and funding)

[6.1] Though Jensen's fatalistic perspective makes his voluminous works very difficult to plod through, one author (Peter) commends Jensen for his effort to get many readers to look squarely at the ubiquitous presence and influence of violence in modern society. If more of the 'well-to-do' of us fully considered how violent our lives are (and how much violence our lives require), we might work harder to reduce it and seek real, collective solutions.

through Family Planning (FP) programs worldwide emphasized controlling population through the control of the female body. Many of these efforts should be considered very successful if measured in their ability to reduce population growth rates (PGRs; see Table 6.1 for the reductions in PGRs over the past 56 years for the 25 most populated countries) and their ability to bring education and prenatal care to rural areas of the global South. Nevertheless, many of the central African nations actually show increases in PGRs due to continued economic hardship, high infant mortality rates and high prevalence of infectious diseases (which counterintuitively increases birth rates). However, some incarnations of these FP programs have done significant short- and long-term physical and psychological damage within many economically-challenged communities worldwide. These effects are well-documented. Millions of women have been forcefully sterilized in the name of 'progress', and millions of others have been subjected to debilitating injections and surgeries as well for similar reasons (SEJUP, 1994; Biswas, 2014); see Kristof and WuDunn's *Half the Sky* (2010) and Hvistendahl (2012) for more explicit accounts.

The focus on population rather than educational and economic inequality masks the great harm that the latter has on populations, 'rich' and 'poor'. When people, particularly women and children, are relatively impoverished, everyone suffers. In Chapter 1, we showed how, below a minimum level of GDP/Capita (see Fig. 1.1), national life expectancy suffers dearly (indicating that many people within such nations are not able to self-actualize and thrive). This isn't the result of 'overpopulation' as it happens in countries of all sizes and densities. As pointed out in Chapter 4, in relation to the issue of life expectancy and energy consumption per person, inequality within a country is a significant variable. When people do not have resources that others have they often behave in ways that are not 'rational' to those that 'have'. The fact that, on average, women from poorer, worse educated backgrounds will have more children than women from wealthier, better educated backgrounds, speaks volumes to this point; this relationship holds for developing as well as developed countries (Skirbekk, 2008). However, as simple as this observation is, it is rarely delved into or properly understood by population-reductionists and many mainstream environmentalists. Obviously, the lack of resources and, in particular, access to basic medical care and secondary education, drives

Table 6.1. Population Growth Rate (PGR) from 1960 to 2016 (most populated nations, 2016)

Country	**Population 2017 (millions)**	**PGR 1960 (%)**	**PGR 2016 (%)**	**PGR Difference 2016–1960 (%)**
"Wealthier" Countries				
US	326	1.7	0.7	–1.0
Russia	143	1.5	0.2	–1.3
Japan	126	0.9	–0.1	–1.0
Germany	81	0.8	1.2	0.4
UK	66	0.3	0.8	0.5
France	65	1.2	0.4	–0.8
Italy	60	2.0	–0.2	–2.2
"Poorer" Countries				
China	1,388	1.8	0.5	–1.3
India	1,342	1.9	1.1	–0.8
Indonesia	264	2.6	1.1	–1.5
Brazil	211	2.9	0.8	–2.1
Pakistan	196	2.3	2.0	–0.3
Nigeria	192	2.0	2.6	0.6
Bangladesh	165	2.8	1.1	–1.7
Mexico	130	3.2	1.3	–1.9
Ethiopia	104	2.2	2.5	0.3
Philippines	103	3.3	1.6	–1.7
Vietnam	95	1.6	1.1	–0.5
Egypt	95	2.8	2.0	–0.8
Congo (Dam. Rep)	82	2.5	3.3	0.8
Iran	81	2.6	1.1	–1.5
Turkey	80	2.4	1.6	–0.8
Thailand	68	3.0	0.3	–2.7
Tanzania	57	2.9	3.1	0.2
South Africa	55	2.5	1.6	–0.9

Source: Population totals: WPR (2017); PGR rates: World Bank (2016b).

young women to have children who cannot be supported economically. Failure to recognize this fundamental point by many, intentional or not, raises the question whether they willingly wish to oppress women (and the communities they live in) in the effort to 'help' them.

And while these oversights by the reproduction-centered environmental movement clearly indicate huge gaps in understanding and/or deliberate patriarchal chauvinism, perhaps the greatest evidence indicating the extreme myopia is in the area of 'missing women'. As a result of many existing patriarchal traditions and societal structures, which privilege the survival of boys over girls, our human population is now ~100 million short of females (Sen, 1993).[6.2] That is right, over the past few generations human actions have resulted in the early death (usually prenatal or early in life) of huge numbers of females; it is a statistic that very few people have ever come across, and you are unlikely going to find much mention of it by the population-reductionists. This horrifying number rivals the total number of deaths associated with 20th Century wars and is approximately double the number of deaths attributable to World War II. Yet, how can we so clearly

[6.2] One hundred million missing women sounds like a very large number, but how could we just not realize this on a day-to-day basis? Consider this thought experiment. Let's say we are at a party that has 200 people in attendance. Now let's assume that these 200 people are distributed gender-wise along the lines of what current global sex ratios dictate (approximately 1.01 males for each female). Guess how many men and women are at this party? There should be 100.5 men and 99.5 women. So, technically, we would not notice this difference, would we? No, we would not and we do not (in our real-world experiences). In a 'healthy' society where all people get adequate healthcare and nutritious food from before the time of birth to old age, the sex ratio is expected to be approximately 0.97 (or 0.97 males for each female). So, if we had representation along these proportions at our fictitious party, we would have 101.5 women and 98.5 men, also something we would not notice. Hence, we see how it is so easy to not see at the micro level what is going on at the macro level and, also, why it is so important to consider how 'seemingly' small differences in societal preferences/practices play out for the entire human population.

Such small differences can be seen in other areas such as climate change, where humans have effectively added another dot's worth of greenhouse gases (GHG) on a landscape that has three dots of GHGs in an atmospheric landscape of 10,000 other dots. Imagine a sheet of paper with 10,000 dots, all black except for three green ones. Would you notice if another green one was added? That level of change is what we humans have added (in the form of CO_2) to the atmosphere since 1800; an almost imperceptible change from certain vantage points but certainly not insignificant to the global climate.

ignore the cultural phenomenon of privileging males when it so clearly has such a grotesque impact? The inability to even acknowledge this fact by major cultures of our modern 'civilized' society suggest how far we must still grow to move forward on the challenges that face all of us.

While these 100 million 'missing' women may not have been documented (like many casualties from foreign wars), they still have been systematically removed from our planet. And since so few mainstream environmentalists even acknowledge the gravity of the absence of these females nor explore the systemic factors that lead to such a horrific outcome, one is left to assume that they are equally blind to the implications of adding more sex-preferential policies (such as the 'one-child' policy in mainland China, operating from 1978 to 2015 when it began to be phased out). Hvistendahl's (2011) *Unnatural Selection: Choosing Boys Over Girls* provides a recent attempt to showcase the extreme consequences of ignoring this issue but this work has generally fallen outside the environmental discourse for reasons that are not so clear. Kristof and WuDunn's (2008) *Half The Sky* offers very compelling examples of the continued horrors that women face today, particularly around sex trade, sexual mutilation, and economical and political oppression. But once again, the connection of these atrocities to broader environmental challenges goes largely unexamined. Continuing this blind attitude in the face of human rights violations so profound bespeaks of ignorance or malice. Obviously, a more holistic and systemic approach to our common challenges seems much preferable. Those who continue to speak of overpopulation as a key problem facing humanity only misrepresent a symptom for a more oppressive disease and in so doing make a grave mistake.

Additionally, the cries about 'overpopulation' usually ignore the rather rapid declines in PGRs that have occurred in the past 40 years. Whereas PGRs were above 2% worldwide in the 1960s, provoking the resurgence of neo-Malthusian ideas, current rates have dropped by more than half, to 1.1% in 2017. The UN currently projects that global populations will level off by 2100 at 11.2 billion (a further increase of 47% on the 7.6 billion living now), though it recognized that the peak could be below ten billion if humane actions are taken to reduce it (United Nations, 2015). However, these projections fail to take into account the bifurcation we have previously noted. If a prevention program to avoid C3 is not successful, global population size may be stabilized or even reduced by mass starvation and epidemics. On the other hand, if such a program is implemented in time, stabilization of global population size may occur sooner and at levels below the UN projection due to diminished economic and social hardship.

6.6. Other obstacles and misdirections

The 'elephant in the room' in this discourse about population and earlier ones on the MIC is, of course, patriarchy, its historical dominance as well as its current ubiquity. Hyper-masculine memes and paradigms dominate modern societies all over the planet. The destructive impacts of these mentalities, and the policies and actions resulting from them, are colossal. The fact that women still make only 70 cents on the dollar of a man in the richest country in the world speaks volumes of the economic oppression that permeates our institutions. As such, no rational, informed discussion of the future can be had without exposing the overly influential role that men have played in structuring societies as they largely now exist. The traps presented by population reductionists, pro-nuclear transitioners (with nuclear's secrecy and overwhelming concentration of power), and MIC boosters, should be clear. In a world where only two national legislative bodies are majority women (Rwanda and Bolivia; IPU, 2017), we must acknowledge the importance gender equality must play in prefiguring discussion of the matters contained in this book and any potential solutions. To fail to do so is to perpetuate the myopia of sexual neutrality. For an illuminating look at the need for more feminine input into any relationship we might have with each other or the natural world, see Patsy Hallen's 1987 piece, 'Making Peace with Nature: Why Ecology Needs Feminism.'

Tracking closely with these sexist, male-chauvinistic perspectives, we find racist ideologies permeating our social structures. Recent events in the U.S. (e.g., the 2016 presidential election, episodes of racially motivated violence perpetrated by police and security officers, predatory lending, etc.) only serve to highlight the pervasiveness of these vestiges of slavery and racial inequality still found in modern society. Acknowledging that racism still exists is important but central to our concerns is how it must be tackled if any effort towards a sustainable society will be successful. An important divide among progressives in the U.S. right now exists between advocates for social causes (such as, addressing police violence, economic hardship, and health challenges) and environmental ones (such as, climate change, biodiversity loss, and the proliferation of toxins). Little progress in either movement will occur until this divide is reconciled. But it should be pointed out that the U.S. environmental justice movement led by Black and Brown people is an already existing critical bridge. Clearly, both groups have a lot to learn from each other and both their issues have incredible overlaps and intersections. Their recognition should generate collective synergisms.

Lastly, many claim that we just don't have enough 'stuff' (resources) to provide for the future human population. The 30-year update of 'The Limits to Growth' illustrates this view with this warning:

> 'Materials are another finite resource. If population rises, and if those people are to have housing, health services, education, cars, refrigerators, and televisions, they will need steel, concrete, copper, aluminum, plastic, and many other materials.
>
> But if an eventual nine billion people on Earth all consumed materials at the rate of the average American, world steel production would need to rise by a factor of five, copper by a factor of eight, and aluminum by a factor of nine. From source to sink, the processing, fabricating, handling, and use of materials leaves a trail of pollution' (Meadows *et al.*, 2002).

We acknowledge that if the world were to consume materials at the rate that the average American does, especially without efficient recycling, we will be facing a global collapse of civilization. However, this assumes that we keep doing things the way we are doing them now — where vast resources are expended on short-lived junk and then converted into trash. Annie Leonard has estimated that the uphill (production-side) pollution stream is 70 plus times that of the downhill (consumption-side) stream. As we have argued earlier, the energy use *per capita* of the wealthy countries of the world will need to come down and the energy use per capita of the poorer nations, and their people, will need to increase. Thus, surprisingly, we actually agree with the Limits to Growth update when it continues with:

> 'Such materials flows are neither possible nor necessary. Fortunately, growth in materials consumption has slowed, and the prospects for

> further slowing are good. The possibilities for recycling, greater efficiency, increased product lifetime, and source reduction in the world of materials are exciting' (Meadows *et al.*, 2002).

This report goes on to say that while the possibilities exist, the actions and directions taken do not make for a promising future — there are just too many people using too many things. We hold that most resources are limited in the sense that we cannot access them because we do not have the energy to do so. Water is a good example. There is obviously enough water on planet Earth for the present population and even more people than the UN projects by 2100, but the vast majority of it happens to be very salty. Well, solar panels can be used to desalinate briny water (as is done in some parts of Saudi Arabia and Chile now). So, as long as we have enough energy we can ensure that we have potable water. Most other resources fit this bill, with solar energy being used to operate a fully recyclable/compostable-based economy. For those resources that are truly limited, such as rare-earth metals and other scarce elements that are currently used by solar energy technologies, they will need to be carefully used and reused, as other, less rare alternatives become viable. Undoubtedly, if humans put significant energy into building biomimetic solar panels out of organic material (mimicking photosynthesis of plants, but at much higher efficiencies of solar capture), humanity would develop energy forms whose abundance is virtually limitless (see Benyus (1997) for a great introduction to how this type of solar energy capture might work).

In conclusion, we have identified several key obstacles to fully contemplating systematic solutions to our many existing challenges. The MIC must be understood for its scale and structural hegemony, most of all for the huge obstacle it poses to implementing these solutions in time. Nuclear and biofuels are not viable alternatives to our fossil fuel dominated system nor reasonable transition energy supplies. Population reductionism has many oversights that expose its irrationality. And, all of these misdirections suffer from highly masculinized roots.

In the next chapter, we will look closely at another giant misdirection, the perspective, held by many environmentalists, that we must uncritically 'degrow' the economy in order to thrive.

Chapter 7

Grow or Die? A Strategy Forward*

7.1. Is degrowth a path to the OWSP?; a critique of degrowth and its politics

We are very critical of the Degrowth discourse, as mentioned at the end of Chapter 4. Here is a representative sample of the Degrowth program:

> 'Perpetual economic growth is neither possible nor desirable. Growth, especially in wealthy nations, is already causing more problems than it solves…Recession isn't sustainable or healthy either. The positive, sustainable alternative is a steady state economy' (CASSE, 2017).

The organizations and websites promoting Degrowth are multiplying faster than the actual growth of the Gross National Product (GNP) of the global economy including: Club for Degrowth, International Conference on Degrowth in the Americas, Third International Conference on Degrowth, Ecological Sustainability and Social Equity, Kick It Over Manifesto, Club of Rome, The Extraenvironmentalist Episode on Degrowth, Degrowthpedia, DéMagazine, Choose Your Future: A Vision of a Sustainable America in 2100, The Path to Degrowth in Overdeveloped Countries, Degrowing Our Way to Genuine Progress, Tim Jackson's *Prosperity Without Growth* (2009), Research & Degrowth (R&D), New Economics Foundation, New Economics Institute, Post-growth Institute, Growth Busters, Growth Bias Busted,

---

*We note that much of this chapter is derived, but revised and updated, from Schwartzman (2011; 2012; 2014d).

Center for the Advancement of the Steady State Economy, Earth Economics, and Sustainable Europe Research Institute.

We will now critique a leading proponent of Degrowth, co-founder of the Italian Degrowth Association, and scholar of ecological economics (EE), Mauro Bonaiuti, because his arguments for Degrowth are illuminating yet misguided (see as well our previous discussion of the views of Nicholas Georgescu-Roegen, the founder of EE, in Chapter 2). In a significant paper promoting Degrowth, Bonaiuti provides an interesting discussion of the social limits of economic growth driven by capital reproduction (Bonaiuti, 2012). Nevertheless, we find the paper's Degrowth argument shallow and frankly incapable of providing a viable political agenda for confronting the converging economic and ecological crises of real existing fossil fuel/nuclear capitalism in the face of the growing threat of catastrophic climate change (C3).

Specifically, like most of what is invoked in a Degrowth approach, Bonaiuti (2012) fails to come to terms with qualitative versus quantitative aspects of economic growth, in particular the critical difference between using current energy supplies to run society versus using a solarized infrastructure to do so. Fundamentally, the concept of economic growth should be deconstructed, with in-depth consideration of its *qualitative versus quantitative* aspects, particularly its differential ecological and health impacts. Growth of what are we speaking? Weapons of mass destruction (WMDs), unnecessary commodities produced because of planned obsolescence, sport utility vehicles (SUVs) versus bicycles, culture, information, pollution, pornography, or simply more hot air? What growth is sustainable in the context of biodiversity preservation and human health, and which is not? Bonaiuti fails to confront these questions and instead lumps all growth into a homogenous outcome of the physical and political economy.

Most of Bonaiuti's paper is occupied with an examination of the social limits of economic growth. However, it fails to account for the highly contradictory character of corporate-driven globalization. For example, now for the first time in history, a majority of humanity lives in urban areas. Population density in the global South has grown alongside great inequalities, while in contrast, a vast new terrain of class struggle has emerged centered on the potentials of green urbanism in the context of the growing threat of C3 (Davis, 2010). Davis' argument for a green (and red) approach to urban reconstruction is very welcome. Even now cities like New York are much more energy efficient than suburbia. China is now the greatest carbon emitter on the planet (not per capita, of course), with more than half its population being urban (55.6% in 2015; CIA, 2017), with urbanization rapidly increasing. Davis points out that very significant reductions in carbon emissions could potentially occur with aggressive energy conversion in buildings and transportation centered in and around urban areas. So, the urban question as a nexus of class struggle is now also a climate security challenge. This huge challenge is also a huge opportunity to create that other world that is possible, especially in metropolitan areas. Undoubtedly, this is a vision that will attract many more adherents than the 'end of growth, we must all sacrifice' mantra of so many neo-Malthusian greens. In contrast, clean air and clean water, meaningful employment, and more free, creative time for all on this planet should be the transnational 'red and green' ecosocialist program. One more example of Degrowth's limited vision should suffice: its failure to recognize that economic growth has created globalized information technology with its immense potential for a transnational movement, a development made possible by this revolution in communication.

Many of Degrowth proponents' arguments with respect to growth in the physical economy and its ecological impacts explicitly rest on the 'bioeconomic criticism' of Georgescu-Roegen. Apparently, they either ignore or are unaware of the published critiques of Georgescu-Roegen and his followers (Schwartzman 1996; 2008; 2009a; 2009b, and references cited within), specifically regarding the fallacious basis of the so-called fourth law of thermodynamics reworded in Bonaiuti's paper (2012) (see Chapter 2 for a detailed discussion).

We recognize that Gross Domestic Product (GDP) is a metric of unsustainable growth because the ever-expanding scale of capital reproduction in the actually existing global capitalist economy is powered by fossil fuel/nuclear/biofuel/big hydropower energy. The imperative of converting the Military Industrial Fossil Fuel Nuclear Terror Complex (MIC) into a high-efficiency solar/agroecological economy will certainly entail real economic growth, as even measured by conventional indices, with a potential of employing virtually all the unemployed on this planet. Thus, we submit that the call should be: Degrow MIC, grow the new green economy! Only once this project is completed, hopefully by 2050, will it be relevant to talk about a global steady-state economy (Schwartzman, 1996).

But to reemphasize what we addressed in Chapter 4, the main obstacle standing in the way of implementing a prevention program to avoid catastrophic climate change (C3) is the MIC and its imperial agenda, blocking a global regime of cooperation needed for a rapid decline in carbon emissions coupled with accelerated wind/solar power creation. In addition, the MIC and its imperial agenda has wreaked grievous harm on both the environment and people, particularly in the global South where mining industries are centered, such as in Latin America, as documented by Gordon and Webber (2016) for the case of Canadian Imperialism.

As pointed out in Chapter 4, demilitarization will free up colossal resources needed for this transition (the present cost of global military expenditures is close to US$2 trillion). As already noted, the direct and indirect costs of fossil fuels total some US$5 trillion per year according to a recent IMF study (Coady *et al.*, 2015). Hence a Global Green New Deal (GGND) is not only cost-effective but an imperative to prevent the onset of C3 while there is still a window of opportunity to act. Nevertheless, while Degrowth may be a problematic recipe for global restructuring, it should not be dismissed as a useless response to the unsustainable reproduction of capital. A reduction in certain kinds of consumption is imperative, especially in the global North and for elites in the global South, since numerous countries such as the United States are such profligate energy and material wasters.

We submit that arguments for Degrowth should be taken especially seriously insofar as they address economic activities that increase consumption of fossil fuels, especially coal, natural gas and tar sands fuel — the GHG emitters with the biggest warming footprint. Struggles against big projects such as the Medupi South African coal-fired power plant (Bond 2010) and fracking around the world are imperative. Likewise, the growing urban organic farming and solar cooperative movements are inspiring examples of how communities can create sustainable alternatives and manage them on a local level, starting in neighborhoods and urban centers.

However, not all big projects should be opposed! Infrastructure must be repaired and replaced, and the immense contamination of our anthroposphere by industrial and military activities must be cleaned up

— these are our responsibility to future generations. Solarization is already occurring on many scales and should continue at a more rapid tempo in the future. This includes ramping up installation of photovoltaics and solar water heaters on homes and apartment complexes and constructing huge offshore wind turbine installations and concentrated solar power in deserts. Likewise, the struggle for social management should range from the neighborhood to the globe, in varied forms of cooperative and nationalized ownership, enlarging the commons by first constraining then doing away with the rule of capital on our planet once and for all. This should be a central objective of the ecosocialist agenda for class struggle in the 21st Century.

7.2. Degrowth, North and South

The Degrowth movement emerged in the global North. Serge Latouche, a retired French professor of economics, has been a leading proponent of the Degrowth movement in Europe. In a recent article, Latouche (2010) argues that degrowth is the 'only political project capable of renewing the Left' as it provides a 'radical critique of consumption and of development...*ipso facto* a critique of capitalism.' He argues that the Degrowth project is 'not about substituting a good economy, *good* growth or *good* development for a bad one and repainting it green, or social, or equitable, with a stronger or weaker dose of state regulation or hybridization through a logic of the gift and solidarity economy [but] about *exiting* the economy.' We find this argument highly problematic for a number of reasons. First, there is no recognition of the qualitative versus quantitative aspects of growth, nor of the material requirements for a high quality of life, in particular the minimum energy consumption per capita. Second, it is a program available for only a minority of the world's people, even in the global North. Who has the option to 'exit the economy?' At best, we can welcome his Degrowth program as pointing to the creation of local food- and energy-producing cooperatives as complementing and supporting class struggles in the real economy; at worst, it calls for a withdrawal from class struggle, a reprise of the (failed) 1960s 'hippy' commune culture.

In contrast, Martınez-Alier (2012) provides a more fruitful conception of Degrowth by recognizing the reality of energy poverty in the global

South and the need to degrow unsustainable consumption in the global North. He recognizes that:

> 'There are enormous inequities in the world, both between North and South, but also in the South and in the North. Some people use per year 250 GJ (gigajoules) of energy [equivalent to 8.0 kilowatt/person], most of which from oil and gas, other people manage with less than 20 GJ [or 0.6 kilowatt/person], including their food energy and some wood or dried dung for cooking' (Martinez-Alier, 2012, p. 65).

He critiques GDP accounting:

> 'Going beyond GDP accounting means something different from "greening the GDP," or at the other extreme, genuflecting before one single environmental index such as the EF [Ecological Footprint]. It should mean to go into a participatory and deliberative multicriteria assessment of the economy, working with ten or twelve indicators of sociocultural, environmental, and economic performance … "Beyond GDP" should mean to set objectives for the reduction in the use of energy and materials and going beyond the single imperative of economic growth, even when this means to leave some financial debts unpaid' (Martinez-Alier, 2012, p. 63).

Further, on the issue of financial debt, he argues:

> 'Instead of becoming obsessed with economic growth in order to repay the accumulated financial debt and supposedly bring happiness to all, rich countries should (at the very least) change their behavior so as not to add to their ever-increasing ecological debt. A program of moderate economic degrowth (implying a lower social metabolism) in the rich industrial economies is a plausible objective to meet this goal. Furthermore, degrowth activists in the North would likely find willing allies in the EJOs [Environmental Justice Organizations] and their networks in the South that are fighting in ecological distribution conflicts against ecologically unequal exchange and the ecological debt' (Martinez-Alier, 2012, p. 64).

We welcome the Degrowth program's emphasis on local autonomy and struggle (Latouche, 2009). But simply acting on a local or even a national scale is not sufficient. There is growing urgency for a transnational ecosocialist movement, a simple recognition that transnational capital and its military arm are blocking an effective and enforceable climate treaty for rapid decarbonization of global energy sources. The senior author observes that:

> 'Bonaiuti and Latouche critique the capitalist mode of production but are rather vague with respect to what should replace it. For example, Bonaiuti (2012) recommends "other forms of economic and social organization more suitable to the new situation." If we take these alternatives to be roughly equivalent to ecosocialism, then ecosocialist political practice needs a robust theory to transcend capitalism. Ecosocialist theory needs to fully engage the natural, physical, and informational sciences, in particular, climatology, ecology, biogeochemistry, and thermodynamics, as well as take full account of the wisdom derived from the experience of thousands of years of indigenous peoples' agriculture and culture. Unfortunately, degrowth proponents take little notice of these sciences, particularly the real thermodynamics of open systems. These sciences will inform the technologies of renewable energy, green production, and agroecologies, whose infrastructure are to replace the present unsustainable mode. Twenty-first Century socialism will be ecosocialism, or it will simply remain the vision of political sects' (Schwartzman, 2012, p. 123).

7.3. A variant of the Degrowth program: Is zero economic growth necessary to prevent climate catastrophe?

In his critique of Li (2013), the senior author says 'Laibman (2013) wisely critiques teleological prognostications of our global future, with the extremes corresponding to utopia or dystopia' (Schwartzman, 2014d, p. 235). While teleologies should be rejected, taking the lead from Ernst Bloch's seminal vision, concrete utopias must be a critical part of strategic thinking if 'strategy' has any intention of implementation.

The ever-growing threat of catastrophic climate change (C3) to civilization and existing biodiversity is the one threat that is very likely inevitable without the implementation of a near-future effective prevention program, unlike nuclear war and global pandemic, which though are continuing threats to be sure, are not inevitable, even without aggressive efforts for prevention. 'However, the survival of our species itself is not likely at risk, nor is the termination of life on our planet, even in the worst-case scenarios for the 21st Century, noting that Hansen has recently backed away from his warning of a Venus-like runaway greenhouse scenario resulting from burning all fossil fuel reserves (Hansen, 2013, citing Goldblatt and Watson, 2012)' (Schwartzman, 2014d, p. 235). In any case, we must take the threat of C3 very seriously, a recognition that should inform arguments for a different global economy, both physical and political. We find in his provocative paper, Li (2013) believes that:

> 'only an economic system that is able to operate a stably with zero economic growth while meeting the population's basic needs could have any chance to address the climate change crisis, ensure global ecological sustainability, and preserve human civilization....Only with zero economic growth (and if necessary, negative economic growth) can reductions in emission intensity through higher energy efficiency and substitution of carbon-free energies be directly translated into absolute reductions of carbon dioxide emissions' (pp. 37–38).

Further in a footnote, he says:

> 'In the future post-capitalist society, "zero economic growth" required for ecological sustainability refers to *zero growth in society's total material production and consumption* [emphasis added]. Thus, zero economic growth does not need to mean the end of improvement in quality of life or the end of the development of productive forces, to the extent "productive forces" are understood to mean human physical and mental development. *Some Marxist writers believe that future socialism needs to be based on both ecological sustainability and economic growth (Schwartzman, 2008; 2009).* I believe, however, that unlimited

> economic growth is fundamentally incompatible with ecological sustainability, for reasons partly discussed in this paper' (p. 38).

Li maintains that only a socialist society will be capable of meeting human needs as well as 'undertake the massive infrastructure transformation within the relatively short period of time required for rapid reduction of emission intensity', foregoing economic growth (Li, 2013, p. 40). The senior author critiqued Li as follows:

> 'I strongly agree with Li's assessment that continuing current trends driven by business-as-usual capital reproduction will result in C3, I disagree with his formulations on fundamentally two counts. First, there is no account of a theory of transition from capitalism to socialism. Li fails to confront the imperative need to struggle for implementation of a C3 prevention program beginning in capitalist society, which I have argued will simultaneously create an unprecedented path to ecosocialist transition (Schwartzman, 2011). Humanity does not have the option to wait until socialism replaces capitalism on a global scale, given the ever-growing threat of C3. My second concern centers on his definition of a zero-growth economy held necessary to achieve ecological sustainability: "zero growth in society's total material production and consumption." While zero growth as so defined is ultimately plausible in a global sustainable civilization, it is an unwelcome prescription for the immediate challenges posed by the threat of C3 as well the undeniable lack of material consumption enjoyed by the majority of humanity living in the global South, the lack of adequate nutrition, housing, education and provision for health services, but most critically, their state of energy poverty' (Schwartzman, 2014d, pp. 236–237).

We again emphasize the relation between life expectancy and energy consumption per capita by nation, demonstrating that most of humanity is now receiving too little energy to achieve the state-of-the-science life expectancy, as discussed in Chapter 4.

As we argue elsewhere in this book, if robust solarization and the reduction of carbon emissions to the atmosphere begin very soon, humanity will have a chance to implement an effective C3 prevention program. Further, this transition will require an increase in consumption of energy and indeed material production to be supplied by a global wind/solar power infrastructure, with many countries in the global North, especially the U.S., decreasing their wasteful consumption, while most of humanity living in the global South receive a significant increase, thereby reaching the rough minimum necessary for acquiring the highest life expectancy. Further, the senior author points out:

> 'Li fails to confront the qualitative versus quantitative aspects of economic growth as well as the thermodynamic implications of the energy source to society. The energy base of the global physical economy is critical: global wind/solar power will pay its "entropic debt" to space as non-incremental waste heat, unlike its unsustainable alternatives (Schwartzman, 1996; 2008)' (Schwartzman, 2014d, p. 238).

It is imperative to emphasize that:

> 'the ecosocialist growth phase, beginning in capitalism itself, must necessarily have a different quality than capitalist economic growth as measured by the GNP, namely not only requiring global growth in the wind and solar power infrastructure, but also in the agroecological sector, with the termination of fossil fuels/nuclear power, industrial agriculture and GMO, and indeed the Military Industrial Fossil Fuel Nuclear State Terror and Surveillance Complex (MIC) itself, at the core of the reproduction of capital [a subject discussed at length in this book]' (Schwartzman, 2014d, p. 238).

The senior author refers to "sustainable economic growth" entailing solarization, demilitarization and ecosystem repair (Schwartzman, 2009, p. 26). We conclude this section as follows:

> 'Degrowth of the MIC should be coupled with global growth of material production required to create a green physical economy and urban spaces, including the repair of the physical infrastructure, expansion of mass transit, including rail, and of course the generation of wind/solar

energy supply to replace fossil fuels/nuclear power. The latter transition should take place on a global scale, but given the global North's historic responsibility for the threat of C3, transfer of wind/solar capacity to the global South from the global North is imperative. The critical metric for economic growth should be its overall ecological and health impacts with respect to the artificial and natural environments, including of course its carbon footprint, and not simply the level of material production, whether measured by spending or some physical unit such as mass' (Schwartzman, 2014d, p. 238).

7.4. A Global Green New Deal as a strategic goal, opening up the path to OWSP

To have any chance left to prevent climate catastrophe, the climate justice movement needs strategic thinking. It is not enough to recognize that the present global system is unsustainable and that a rapid and radical transformation is necessary. And not only does a strategy require a plausible plan to bring about such change — recognizing the obstacles and how to overcome them — it must also have the capacity, when implemented, to ignite the imagination of millions around the world to form a collective transnational Subject, arguably the only force capable of preventing climate catastrophe and creating the other world still possible (OWSP). We hope our discussion will, in its modest way, help promote such a strategy before humanity plunges into the abyss of climate hell.

We are convinced that a critical component of this strategy must be bottom-up movements of all those bearing the weight of this world of extreme disparity and inequality: the exploited and the oppressed. Prefigurations of the OWSP are being created all over the world — whether they be farming or solar power cooperatives, worker-owned industries, or community land trusts creating affordable housing outside the constraints of capital-driven speculation in urban areas. We reject the dogma that competitive and individualist behavior will always drive social interactions because 'you can't change human nature'. Even today, in real existing capitalist societies, cooperation spontaneously springs up when communities are under attack.

We are not content to simply dream of a different future, smoking the pipe of a drug-enabled escape from present realities. Nevertheless, the process of lucid dreaming, imagining a future utopia, is a necessary place to start. As Ernst Bloch (1986) discusses, the recognition of the 'Not-Yet' is an anticipation founded on the principle of hope, the title of his seminal three volume magnum opus. We provocatively quote Lenin, 'We ought to dream!' (Lenin, 1929, p. 158), notably a passage quoted by Bloch (1986), with the inspiration of 'Imagine' (John Lennon's well-known song). Thus, we submit the outline of a strategy that must begin where we are now, a strategy for a Green New Deal (GND), indeed a Global Green New Deal (GGND), which even with partial realization could significantly improve the quality of life for millions around the world. At the same time, the implementation of a GGND has the potential of opening up a path to a global ecosocialist transition, to create the OWSP. The struggle for a GND must include achieving a real measure of national state power, even entities of transnational state power, synergized with the growing strength of bottom-up prefigurations.

Can we draw lessons from the experience of the success of the New Deal during the Great Depression of the 1930s as we consider a potential Green New Deal approach to dealing with the current deep-rooted and multifaceted crises facing capitalism today? Contrary to commonly held belief, FDR's New Deal was implemented to save capitalism, and its most progressive initiatives only came as a response to fierce class struggle, including the resurgence of the industrial worker movement, which resulted in the formation of the Congress of Industrial Unions in 1936. Note this history:

> 'While the bureaucratic leadership of the AFL was unable to win strikes, three victorious strikes suddenly exploded onto the scene in 1934. These were the Minneapolis Teamsters Strike of 1934, the leadership of which included some members of the Trotskyist Communist League of America; the 1934 West Coast Longshore Strike, the leadership of which included some members of the Communist Party USA; and the 1934 Toledo Auto-Lite Strike led by the American Workers Party. Victorious industrial unions with

> militant leaderships were the catalyst that brought about the rise of the CIO' (Wikipedia, n.d.a.).

Even though manufacturing employment increased by 3 million from 1933 to 1939, the unemployment rate did not significantly decline, staying between 15% and 20% until military-related production started taking off by 1940 as a result of WWII (Wikipedia, n.d.b). The organized Left, socialist and communist, had grown into a relatively strong political force that was only smashed after a concerted anti-democratic campaign, which demonized those who challenged unfettered capitalism, by the defenders of capital during the Cold War. This effort eventually succeeded, culminating with the leadership of organized labor making its Faustian bargain with capital to purge the Left and collaborate with the newly emerged Military Industrial Complex (MIC) and its imperial agenda in exchange for promises that the real wages of workers would rise to unprecedented levels. This worked for a short time, until capital began to renege on the deal with the neoliberal restructuring of the economy that began in the 1970s. Since then, wages for the vast majority of people have not only stagnated, but continued to diverge from increases in labor productivity (Bernstein, 2016).

Now, in the current profound crisis of capitalism driven in part by financialization, we are likely facing a prolonged period of high precarious employment, with significant sectors of the U.S. 'middle' class approaching the insecurity of marginalized workers in the global South. And for the first time since the Great Depression, older workers are facing the prospect of permanent unemployment, while even educated youth now confront a bleak future of part-time work and ever-accumulating debt. These factors very likely contributed to the election of Donald Trump as President of the United States on November 8, 2016, noting the role of voter suppression and the white supremacist Electoral College, given the fact that his main opponent had nearly 3 million more votes nationally than him.

The Green New Deal has been championed as a solution to the current crisis, and as a green Keynesian reprise of the New Deal. It has the potential of generating millions of new jobs both in the energy conservation/clean energy sector and in the repair of physical infrastructure. The

BlueGreen Alliance (2017) in the U.S., One Million Climate Jobs (2017) in the U.K., the Global Green New Deal (GGND),[7.1] and the Pan-European GND (Yaroufakis, 2014) are all examples of GND initiatives.

[7.1] See, for example, the following article from the International Trade Union Confederation (ITUC, 2011) regarding a global green economy. Here is the post in its entirety:

> 'A green economy can mean higher overall employment and better jobs, and is not just a luxury for wealthy countries, reveals the UN Environment Programme (UNEP) in its Green Economy report released today. Supported by concrete examples from around the world, and a thorough macroeconomic analysis, the report underlines what the labour movement has maintained for several years: that a Green Economy, based on the right principles and properly planned, can deliver for workers and the poor.
>
> "I am pleased to read that UNEP shares with workers around the world the deep belief that a green economy should work for the people and the planet, and not just for GDP growth and a few wealthy companies," said ITUC General Secretary Sharan Burrow. "As the report signals, one of the challenges is to ensure a just transition that will steer transformation across all sectors of the economy and lead U.S. towards the decent and sustainable jobs of tomorrow."'
>
> The report indicates that the allocation of 2% of global GDP towards the green economy could lead to immense benefits for workers and communities around the world and help overcome the diverse challenges countries are facing. It finds that a green economy can generate at least as much employment as the traditional economy and outperforms the latter in the medium and long run while yielding significantly more environmental and social benefits. The report stresses the importance of ensuring trade union rights — including freedom of association — and occupational health and safety in traditional and emerging sectors.
>
> To those who consider that there is a high risk of 'green and social-washing', Burrow reacted as such, 'The risk is the status quo'. The green economy presents an opportunity to engage in a transformational path towards sustainable development. We must ensure this is not misused, we must ensure that [the] green economy works for working people.
>
> The UNEP report sets out a clear pathway towards a green economy, but policies being pursued by governments at the moment risk taking us backwards. Neoliberal recipes, based on the dictates of the financial markets, have to be jettisoned in favour of a progressive approach in which governments fulfill their responsibility to regulate banking and finance, promote policies which stimulate greening of workplaces and creation of new green jobs, and ensure that this is based on social dialogue and social inclusion', explained Burrow.

These developments are barely visible in the writings of Marxist critics of Green Capitalism, such as Richard Smith (2011). Although Smith's critique of market-driven Green Capitalism is timely, thorough, and to the point, he concludes that the only real solution is:

> '...collective democratic control over the economy to prioritize the needs of society and the environment. And they require national and international economic planning to re-organize the economy and redeploy labor and resources to these ends. I conclude, therefore, that if humanity is to save itself, we have no choice but to overthrow capitalism and replace it with a democratically planned socialist economy' (Smith, 2011, p. 112).

While we are in general sympathy with Smith's conclusion, he provides not a trace of a strategy to achieve this goal. Will workers have a role in overthrowing capitalism, other than perhaps by some miraculous undescribed process of conversion, we suppose, during this system's terminal illness? According to Smith, apparently workers simply are forced to share the same goals as the capitalists: 'So CEOs, workers, and governments find that they all "need" to maximize growth, overconsumption, even pollution, to destroy their children's tomorrows to hang onto their jobs today because, if they don't, the system falls into crisis, or worse' (Smith, 2011, p. 112). Other than those who will presumably become socialists just by virtue of reading a convincing article, who will be the gravediggers of capitalism?

Along with specific education and training policies to ensure the skill needs for a green economy are met, economic safety nets and social protection are crucial to achieving the necessary transformation in a way which maximizes the economic and social benefits.

A green economy which works for social justice can only be a collective endeavour; it should therefore be equitable, inclusive, democratic and people-centered. We will continue to push the case, in the coming days at the UNEP Governing Council meeting in Nairobi, and also during the next 16 months in the run up to RIO+20, to make the green economy a driver for prosperity and decent work', concluded Burrow.'

Smith (2011) does hint at one narrow window of opportunity under real existing capitalism for socialist intervention, the promotion of non-market solutions:

> '...the only way to prevent overshoot and collapse is to enforce a massive economic contraction in the industrialized economies, retrenching production across a broad range of unnecessary, resource-hogging, wasteful and polluting industries, even virtually shutting down the worst. Yet this option is foreclosed under capitalism because this is not socialism: no one is promising new jobs to unemployed coal miners, oil-drillers, automakers, airline pilots, chemists, plastic junk makers, and others whose jobs would be lost because their industries would have to be retrenched — and unemployed workers don't pay taxes' (Smith, p. 112; identical passage in Smith, 2016, pp. 49–50).

But this triggers the following question: didn't massive job creation for the unemployed, financed by government spending and not the market, actually happen once before in the history of capitalism — i.e., in the New Deal, when organized workers, employed, and unemployed forced it to happen through collective class struggle? Unfortunately, the possibility of class struggle within capitalist society seems to be left out in Smith's account. Didn't class struggle, in its broadest aspects, involving the environmental/occupational health movements, actually win non-market regulatory power in 20th century U.S. capitalism in the form of the Clean Air Act and the establishment of the Environmental Protection Agency and Occupational Safety and Health Administration? To be sure, these agencies have been diluted and weakened with the neoliberal offensive of capital over the last 30 years. And indeed, barely constrained by a very weak regulatory regime, the global expansion of capital reproduction has brought us ever closer to irreversible tipping points to ecocatastrophe.

But is Smith's 'massive economic contraction in the industrialized economies' really required to prevent such an outcome? The qualitative aspects of growth, and not simply the level of the GDP, need to be addressed. And yes, we absolutely need the rapid economic contraction of the military/fossil fuel/nuclear-powered economy that produces the vast amounts of short-lived commodities that either end up in ever-mounting junk piles or are destroyed in resource wars. But what about the economic expansion of the green economy producing clean energy and repairing and restoring the physical infrastructure, including mass transit and green cities (Jones, 2008)? This possibility is ruled out by Smith:

> 'Renewable energy scientists argue that integrated comprehensive systems can solve the problem of base-load generation. The IEA [International Energy Agency] estimates that solar power alone could produce almost a quarter of the world's electricity needs by 2050... But as Ted Trainer points out, given the variable and intermittent output of renewables like solar and wind, even if sun and wind were to be large contributors to electricity supply, given the need for backup reserve capacity, little or no reduction in the amount of coal or nuclear capacity would be feasible' (2011, p. 133).[7.2]

We strongly disagree with this assessment appropriated from Trainer (see P. Schwartzman and D. Schwartzman, 2011; D. Schwartzman and P. Schwartzman, 2013; Schwartzman, 2014b; 2014c). Smith goes on to argue:

> 'Yet even if we could get a dramatic shift to solar and other renewables for energy generation, given the Jevons paradox...we cannot assume that this would necessarily lead to large permanent reductions in overall pollution. For if there are no non-market constraints on production, then the advent of cheap clean energy production could just as easily encourage the production of endless electric vehicles, appliances, lighting, laptops, phones, iPads and new toys we can't even imagine yet...The expanded production of all this stuff, on a global scale, would just consume ever more raw materials, more metals, plastics, rare earths, etc., produce more pollution, destroy more of the environment, and all end up in some landfill somewhere someday. In short, at the end of the day, *the only way society can really put the brakes on overconsumption of electricity is to impose non-market limits on electricity production and consumption,*

[7.2] Note that Smith (2016, p. 81) omits this reference to Trainer even though his Introduction states that this chapter was taken from his 2011 publication.

> *enforce radical conservation, rationing, and stop making all the unnecessary gadgets that demand endless supplies of power'* (2011, p. 134, emphasis in the original; very similar passage in Smith, 2016, p. 82).

Indeed, we recognize that imposing such non-market limits is imperative, but the struggle to so impose them must begin in capitalist societies now, and not be posed simply as the policies of post-capitalist society. And yes, by all means, aggressive energy conservation is necessary, especially in the United States and other countries of the global North. We can all live better with a sharp reduction of wasteful consumption, and breathe clean air, drink clean water, and eat organic food. Nevertheless, there is an imperative for a global increase in the power capacity, employing clean energy and not fossil fuels or nuclear power, to ensure every child born on this planet has the material requirements for the highest quality of life (P. Schwartzman and D. Schwartzman, 2011; D. Schwartzman and P. Schwartzman, 2013).

It should be noted that in his more recent work, Smith (2016) does imply that moving to a sustainable economy should begin under capitalism by listing '*at least some* or all of the following' [*italic* added] steps, including 'put[ting] the brakes on out-of-control growth in the global North — retrench[ing] or shut[ting] down unnecessary, resource-hogging, wasteful, polluting industries like fossil fuels…' (p. 148). Nevertheless, once again, the agent of change is not spelled out.

A welcome critique of Green Capitalism is found in Fitz (2014), focusing on its delusionary promise that business-as-usual market-driven economic growth has the capacity to deliver a truly sustainable future for humanity. But in the end, he fails to recognize both the immense opportunity, indeed the imperative, of supporting a Global Green New Deal (GGND) for any hope of avoiding catastrophic climate change (C3) and ending energy poverty for the majority of humanity living in the global South. And Fitz fails to recognize the necessity of engaging in multidimensional class struggle to shape the content of a GGND as a path opening to ecosocialist transition.

In addition, Fitz writes: 'An increasingly popular answer is the "Green New Deal" (GND): create "green jobs" in order to jump start the economy. But the GND **might not provide** long term employment and **could cause** major environmental harm' [bold added] (Fitz, 2014). While Fitz used the terms 'might not' and 'could', we are of the opinion that these are, in reality, contingencies that must be the nexus of class struggle, not simply possibilities to consider.

We have major disagreements with Fitz's arguments concerning the energy needs for humanity, economic growth and the issue of large scale creation of wind and solar energy infrastructure to replace our present unsustainable supplies. Fitz correctly notes, simply building renewable energy capacity is not sufficient to prevent dangerous climate change, as global carbon emissions continue to climb. However, Fitz calls for reductions in 'total energy usage (not merely fossil fuel); and, total industrial production' (2014). Fitz's opposition to economic growth associated with a GGND is unwelcome because it fails to analyze the quality of such growth, i.e., what needs to grow and what needs to degrow (see previous discussion of the Degrowth movement).

But should we anticipate that Green Capitalism, even pushed to its limits by class struggle, *could indefinitely postpone* the final demise of global capitalism and could actually replace the present unsustainable energy base with a renewable power infrastructure fast enough to avoid catastrophic climate change (C3)? We submit this prospect appears to be highly unlikely because the legacy and political economy of real existing capitalism alone makes *global* solar capitalism a delusion (Schwartzman, 2009a). And at the heart of this legacy is the Military Industrial Complex (MIC) (see discussion in Chapters 4 and 6). It should be noted that confronting proposals for a GND, Green Capitalism and a Green Economy should take into account the important distinctions and nuances embedded in their details, particularly in relation to the prospects of sustainable development in the global South (Tienhaara, 2014).

Nevertheless, what the struggle for a GND can accomplish is very significant, indeed critical, to confronting the challenge of preventing C3. *Humanity cannot afford to wait for a post-capitalist society to replace capitalism to begin implementing this prevention program.* And the senior author has argued (Schwartzman, 2011), starting this prevention program under existing capitalism can open up a path toward ecosocialist transition, indeed a 21st century Socialism worthy of its name. This perspective is endorsed in an analysis of the GND by Asici and Bünül (2012).

Climate science tells us we must proceed *now* for any plausible chance of avoiding tipping points plunging us into C3. Green job creation is likewise the creation of a new working-class sector committed to ending the fossil fuel addiction. Such an historic shift to renewable energy supplies would be comparable to the industrial revolution that replaced plant power in the form of wood and agricultural products with coal. Though this is necessary, it alone will not be sufficient for preventing C3. Given the Jevons paradox, as Smith notes, we will also need to implement a strong regulatory regime for curbing carbon emissions in order to avoid just expanding renewables as a supplement to the continued reliance on fossil fuels. At the same time, a broad global alliance of the working class and oppressed people including blue green labor, women's movements, marginalized workers, indigenous people, and yes, to be sure, factions of capital investing in solar, can potentially create the power to challenge the privileged position of MIC capital and begin the process of global demilitarization, putting the MIC dinosaur in the Museum of Prehistory where he belongs, to be followed soon by his parent, Global Capitalism.

The socialist Chris Williams also advocates including 'solar capital' as a component of this global alliance, when he observes:

> 'Some of the more far-sighted corporations without significant investments in fossil fuels will see the way the wind is blowing and that money can be made from investing in alternative energies, as is already the case. This will create tension and splits among ruling elites and between conflicting corporate interests, which will open up space for social and labor movements to demand swifter and more coordinated action' (2010, p. 166).

Further, we emphasize that the struggle to achieve this broad alliance must overcome the divisions in the working class and its potential allies, which the ruling classes and elites encourage to their advantage. The class struggles in the New Deal era made significant gains in the U.S. in fighting racism, and of course the civil rights movement of the 1940s through the 1960s achieved many victories codified into law. However, we are now faced with the continuing legacy of mass incarceration of mainly minorities, the 'New Jim Crow' (Alexander, 2010), as well as the demonization, exploitation and repression of immigrant communities, especially those from the global South. These new challenges must be confronted by the organizers for a GND, since unity, rather than division among oppressed peoples, will take away a powerful tool that ruling elites use to distract the masses from the abuses they themselves inflict on the majority.

And unlike the New Deal, we expect that achieving the GND on a global scale in the context of a robust solar transition, by necessity accompanied by demilitarization, will not end with a reinforcement of militarized capital, as was the case in WWII and the Cold War aftermath. Rather, the GND has real potential for opening up a path out of capitalism into ecosocialism. WWII and the emergence of the MIC postponed the terminal crisis of capitalism to this century. Now we face the welcomed project of taking that terminal crisis on and finishing the job. We need a strategy of transition. This should be a priority to be considered, both in theory and practice for all those looking for a path out of the profoundly unsustainable situation humanity now faces. Of course, the immediate needs of the great majority of those exploited and oppressed by big capital must be confronted. Therefore, jobs, affordable housing, healthcare and child care, environmental quality, and environmental justice must be high on the agenda for political activity. Likewise, this agenda must confront the ecological crisis and demand solutions that address climate change by embracing clean energy. But what kind of jobs? For unsustainable or sustainable green production?

We should never advocate or even imagine that the 'worse the better' will deliver a utopia by the collapse of capitalism, anticipating its terminal illness as hope, for capitalism's dead weight will kill us all. No slogan or propaganda alone can achieve success, as important as this ideological struggle is. Rather, only multidimensional and local-to-transnational class struggle within capitalism (see e.g., Abramsky's illuminating account, 2010) can terminate this system, which unfortunately will not die a natural death on its own accord. It will have to be put to sleep forever. *A critical role of the ecosocialist Left is to identify the strategic class sector — those existing and those in formation — that will be the gravediggers of capitalism.* Additionally, the ecosocialist Left must also, of course, participate in the creation of a collective vision and its realization as embryos within capitalism culminating into a new global civilization ending the rule of capital.

And among these 'gravediggers', more positively the critical component of the global Subject that will create the OWSP, are workers organized in trade unions, especially in the growing renewable energy and information technology sector. The following is representative of the cutting-edge thinking of this sector:

> 'Trade Unions for Energy Democracy (TUED) is a global, multi-sector initiative to advance democratic direction and control of energy in a way that promotes solutions to the climate crisis, energy poverty, the degradation of both land and people, and responds to the attacks on workers' rights and protections…We are facing an energy and climate emergency that amounts to a planetary crisis. The growing levels of fossil-based energy is stretching planetary limits by raising greenhouse gas emissions and air pollution to alarming levels. This is affecting the health and quality of life of millions...The power of fossil fuel corporations has made it practically impossible to protect the health and safety of workers and communities, and union representation is under attack across the globe. Despite more energy being generated every year, energy poverty remains a serious global issue — 1.6 billion people, or 20% of the world's population, do not have regular access to electricity. It has become increasingly clear that the transition to an equitable, sustainable energy system can only occur if there is decisive shift in power towards workers, communities and the public. The goals of the project are:
>
> Help build and strengthen a global trade union community for energy democracy. TUED is a platform for trade unions from all sectors and countries to debate, develop and promote real solutions to the climate crisis, land grabs, energy poverty, and pollution generated by fossil fuels — solutions that can build unions, worker and community power, and advance social and environmental justice.
>
> Develop high-impact union educational materials, distribute an electronic bulletin, and convene meetings and working retreats that encourage debate and help create a shared analysis of key energy and climate issues.
>
> Connect the energy democracy agenda to union struggles and campaigns in ways that build broad membership engagement, increase

worker power, and facilitate solidarity across movements that share similar goals' (Trade Unions for Energy Democracy, 2017).

A thorough analysis of the role of labor and trade unions with respect to the creation of a GGND and the challenge of climate change is found in Barca (2015). Kunze and Becker (2015) discuss the role of collective ownership of renewable energy projects, particularly on a small-scale in the European Union in the context of the Degrowth concept.

We now witness or can soon anticipate ongoing struggles for social governance of production and consumption on all scales from neighborhood to global. Areas of struggle in this fight should include: the nationalization of the energy, rail, and telecommunications industries; municipalization of electric and water supplies; the creation and maintenance of decentralized solar power, food, energy and farming cooperatives; the encouragement of worker-owned factories (the solidarity economy); the replacement of industrial and GMO agriculture with agroecologies; the creation of green cities; and, of course, the organization of the unorganized in all sectors, especially GND workers. All of these objectives should be part of the ecosocialist agenda for struggles around a GND, which of course, must include the termination of the MIC. One outstanding example of how to begin is found in Davis (2010), who argues for the potential of a radical movement for green urbanism (Schwartzman, 2010).

The GND entered the U.S. political discourse during the 2012 and 2016 presidential elections as a result of its focus in the platform of Dr. Jill Stein, representing the Green Party of the United States (Stein, 2012; Green Party of the United States, 2016). Likewise, a GND for New York State was highlighted in Howie Hawkins' Green Party candidacy for governor in 2014. The experience of attempts to reconvert military production should be studied closely in this regard, since demilitarization should be an essential component of the GND (e.g., Feldman, 1991; 2006; 2007). A GND is essential to make possible a robust transition to a global renewable energy infrastructure (FitzRoy, 2017).

The history of capitalism is the history of class struggle, its ebbs and flows. It is certainly not a history of the working class as a passive instrument, a lever in the machine of capital reproduction. To write off class struggle is to revert to the empty idealist prescriptions of what ought to be rather than focusing on materialist theory and practice to make it happen, beginning within the womb of present society. But this theory and practice must also be informed by a spiritual dimension, the anticipations of what can be realized in the future, concretized in prefigurations inspired by the principle of hope. The potential of a Green New Deal, starting as soon as possible, gives us a powerful wedge if we choose to use it.

7.5. Conclusion

The Degrowth program is highly problematic because of its failure to analyze the qualitative aspects of economic growth and its emphasis on the local economy without recognizing the urgency to address global anthropogenic change from a transnational political perspective. A largely absent focus in the Degrowth program is the major challenge humanity now is facing: to rapidly implement a prevention program to avoid catastrophic climate change. A rapid transition entailing demilitarization, solarization, and the creation of agroecologies is critical given the ever-narrowing window of opportunity. This demands struggle on all spatial scales, from the neighborhood to the globe.

Chapter 8

The Revolution has Begun

8.1. Introduction

Excitement reigns upon reflection of all the opportunities available to us. Given the media's obsession with the grotesque, the fear-inducing, and the entertaining, many have not even considered what is really attainable. As such, the few that offer ideas for a better future are received as 'dreamers' and 'pie in the sky' types and generally ignored; authors have experienced this in their own communities. No doubt, as shown in Chapter 6, there are many obstacles that challenge a natural transition to the other world still possible (OWSP). However, an examination of some of the activities taking place right now confirms that the beginning of a transition/revolution is actually currently well under way. Thus, by expanding these efforts beyond their current artificial boundaries, we will bring about the transition/revolution, with all its benefits. Undoubtedly, the future success of humanity depends on understanding what makes these activities thrive and implementing them thoughtfully — bringing them to scale while tailoring them to each community's particular needs and circumstances. We must be inspired by the efforts currently afoot. With this energy and enthusiasm in mind (and heart), we can collectively begin the difficult, yet impactful, rewarding work ahead of us.

8.2. The future of food in motion

Literally thousands of communities in the U.S. and elsewhere have been engaging in or otherwise supporting sustainable forms of agriculture. Cobb's (2011) *Reclaiming Our Food: How the Grassroots Food Movement*

Is Changing the Way We Eat does a spectacular job of covering a multitude of these efforts in seven different regions of the United States. From Rose Marie Williams teaching Navajo heritage food and agriculture in northern Arizona, to the Appalachian Sustainable Agriculture Project assisting in the connecting of growers, purveyors, and consumers in North Carolina, the seeds for a sustainable agricultural system (pardon the obvious pun) are being planted each and every day. And, notice that Cobb's book is more than half a decade old. Some projects, such as Gateway Greening (in St. Louis) and P-Patch (in Seattle), while others, such a high tunnel erected in a town in Illinois or a burgeoning basement mushroom grower in rural Connecticut, still go without much notice. However, all are indicative of a movement that continues to expand, often 'under the radar'.

Several quantifiable indicators provide demonstrable evidence that the 'New Ag[riculture]' movement is vast and growing quickly. Thousands of small 'organic' farms (many do not go through the tedious bureaucratic process for official status, as they operate organically) and vegetarian/vegan restaurants have opened up all over the U.S. Many colleges (and high schools) are offering courses in sustainable or urban agriculture. These activities demonstrate that growth is society-wide and not just in one sector. Another key indicator of the interest in non-industrial forms of agriculture is the growth in number of farmers' markets in the United States. Remarkably, from 1994–2014, they expanded in number from some 1,800 to about 8,300, a growth of 360%, though, apparently, the rate of growth has recently begun to decline (Leibrock, 2014).

While this growth is promising and indicative of a flourishing movement, 'sustainable' food still represents a small dent of all food consumed in the United States. Organic food now accounts for some 4% of all food sold in the U.S., though the total sales of such food nearly tripled from 2005 to 2014 (USDA, 2017c). Major changes in agricultural policy will clearly need to occur to increase this percentage substantially, given the fact that a few grain crops are still being produced in massive amounts, mostly geared towards meat, fructose, and ethanol production. Academics can contribute to this movement by teaching food/agriculture/resource courses that motivate students (i.e., potential future leaders and importantly farmers) to understand that real alternatives exist. Consumers can play a major role by purchasing locally grown and organically-raised food whenever possible. But as large numbers of people cannot financially afford the additional expenses that this purchasing shift requires, we recognize that this last recommendation has limited application. As academic and consumer efforts will be insufficient to bring about the changes necessary, something grander must take place. It will take many seeds of 'change' to foster the scale of modifications necessary — industrial agriculture remains a ubiquitous and powerful entity. Fortunately, seeds are found almost anywhere one looks these days. In what follows, we document what we found when we traveled some about the terrain of the United States and Cuba in the spring and early summer of 2017. It represents an important snapshot of what is happening and what could happen elsewhere with greater commitment.

8.3. Detroit's food path

Detroit, Michigan is one place where innovative food stuff is happening. Historically, Detroit was once the fifth-largest city in the U.S. (in 1950), when it was home to nearly 1.9 million residents. Today's Detroit, for a variety of reasons (e.g., companies moving overseas and white flight to suburbia post WWII), has become the eighteenth largest city (home to fewer than 700,000 in 2013, corresponding to a 65% reduction in population in about 65 years). When Mayor Coleman Young (the city's first Black mayor) took office in 1973, outsourcing and deindustrialization was well entrenched and new directions appeared necessary. Sadly, Young capitulated to the big developers and endorsed the razing of Poletown (displacing 1,300 households) and supported the 'casinoization' of the area (Boggs, 2012). These efforts further expanded zones of poverty and blight. In 2007 the last supermarket in Detroit was closed (Greenbaum, 2014). However, out of the literal rubble that continued to manifest in formerly thriving neighborhoods, which we both saw with our own eyes in June 2010, during the U.S. Social Forum held in Detroit, residents started to develop autonomous enterprises as a means of survival.

An early summer 2017 field trip to Detroit revealed several agroecological efforts active in the 'Motor City'. Food Field, founded in 2010, produces herbs, vegetables, fish, chicken and duck eggs, and tree fruits on a four-acre lot (formerly occupied by a public school). An interview with its founder, Noah Link, elicited this assessment: steady, yet slow, growth in sales, particularly through an expanding CSA (Community Supported Agriculture program) and significant unmet demands from some local restaurants suggesting more and more food purveyors are looking to local, fresh, and chemical-free foods.

D-Town's Farm, a seven-acre farm on the far northwest side of Detroit founded in 2006 recently expanded its acreage and erected another high tunnel, solar panels to run its motorized equipment, and, soon, an early season seedling operation. Its farm stand sees increased traffic and its annual festival continues to draw significant interest as well. But, D-Town's farm isn't just about growing crops. They use farming as a 'community-based resistance strategy with a political change initiative' in recognition of the larger struggles facing communities everywhere, particularly communities of color (Greenbaum, 2014). Food Field and D-Town Farm have also integrated PV panels, wind turbines, and even emus to make their farms flourish, see Fig. 8.1. Fresh Cut Detroit, in the

Figure 8.1. Detroit farms: (top) D-Town: solar panels (used to heat high tunnels) (right) D-Town: wind turbine (used to pump water from retaining pond to plants) (left) Food Field: Solar for aquaponic system (in high tunnel) & emus to protect chickens (Photos: Peter Schwartzman)

heart of the City, operates a flower farm (with more than 200 varieties) and has recently seen expansion in land and sales as well. Sarah, its founder, grows 'in organic settings [and bases] decisions on a balance between organic standards and research, long-term health of land, workers and customers, and productivity balanced with fertility' (Steph, 2014). The Capuchin Earthworks Farm (CEF) east of Downtown Detroit operates two high tunnels on 2.5 acres and is certified organic (CSK, 2017). It also operates a soup kitchen on site and a weekly market stand. Perhaps most importantly, CEF 'grows transplants for thousands of community gardens' throughout Detroit as well as provides educational programming for the James and Grace Lee Boggs School (named after the very influential global human rights and peace advocates who spent many years in Detroit fighting for justice).

While impressive, these projects just touch the surface of all the urban farms and food-orientated organizations and the many people associated with each of them in Detroit. Apparently, there are over 1,000 community gardens in Detroit, according to Keep Detroit Growing, a key organization which also supports innumerable local farm operations through education, startup equipment, seeds, and, importantly, soil tests (particularly for lead contamination) (Detroit Ag, 2017). And last but not least, Detroit's large Eastern Market, open on Tuesdays and Saturdays, claims to be one of (if not) the biggest and oldest year-round farmers' markets in the United States. Though not all local farms participate in the Eastern Market (as many host their own markets, enabling the food to be even closer and more accessible to its neighbors), a trip to the Market revealed hundreds of vendors — nearly all of whom sold local food and plants (to fuel other small grower's production efforts and/or enhance ecological diversity in the region). Importantly, several of the vendors at the Market were larger, sometimes organic, farms located in rural areas outside of Detroit. Dining out revealed that many local restaurants are purchasing produce (and value-added products) from local farms as well.

Though these projects are found all over Detroit (one of the top 70 largest cities in the U.S. by area, encompassing 139 square miles) and each has something unique about its efforts, they all share a few common themes that are important as we begin to understand them as the *prefiguration* of the OWSP. For one, they are dedicated to growing in ways that complement nature rather than wage war against it. All the farms use little if any pesticides, and most retain significant natural habitats in/around the farm. This decision is a conscious one and has several beneficial outcomes. It enables biodiversity to thrive. It also provides for natural shade that benefits the workers and visitors during oppressively sunny days. It also reduces erosion. The farms are set up to provide real employment to local people, though the specifics on salaries, or benefits, were not readily available. The farms all had educational components — via tours, internships, on-site social events, or training sessions (onsite or offsite). Some of these services were done as part of their core institutional mission, but others were essential, either in order to take advantage of available volunteer labor, or as insurance for maintaining productivity on the farm through reproducing knowledge and continuity. Importantly, all the food farms are focused on getting food to local residents. Obviously, the farms must sell food to provide sustenance (via income) to the farmers. However, each farm appeared to recognize that a significant portion of Detroit's population has challenges getting sufficient food, particularly fresh, nutritious food. As such, many farms appeared to sacrifice some of their profits (by selling produce at reduced prices) when selling on site (to residents of their neighborhood). This represents just one more way these farmers fully appreciate their place/space in the larger society of which they are part. This type of engagement raises all ships at once.

Detroit's urban agriculture efforts have been impressive and substantial but challenges still exist to make it flourish. These challenges are not unique to Detroit as many are found in other metropolises. First and foremost, small farmers don't have dependable income, and even when they do have money coming in it almost always tends to be seasonal. Obtaining reliable customers and entry into successful markets remains difficult for many small farms. Given the important contribution that these farmers are making to society, we need to find ways to pay them all a living wage, year-round, as well as provide them with healthcare. Efforts to ensure that 'food stamps' and similar government subsidized food assistance programs are usable at Farmers' Markets and CSAs can help bring sizable resources to local farmers as well. Subsidies for institutional purchases of local food (at hospitals, schools and prisons) could also inject considerable dollars into local economies. Large scale farmers have some protections from 'bad years' but these small urban farmers have very few. And since many of these small farms are started by individuals who may not be well-versed in business or organizational development, there is a steep initial learning curve. These forward-thinking entrepreneurs, working on their own, already have difficulty making ends meet, just imagine the difficulty they confront when hiring others becomes necessary, as it almost always does. On small farms, there are many tasks to be done and it is very difficult for one person to do everything. Economies of scale definitely benefit the success of these farms. Efforts launched by Keep Growing Detroit (and SAAFON, Southeastern African American Organic Farmers Network, headquartered in Georgia) are invaluable to these small farms as they assist them organizationally during their early growing stages.

Cooperation Jackson (which we visited in March 2017) is a very exciting attempt to organize urban agriculture, cooperative markets, technological innovation (e.g., 3-D printing), youth mentoring, and social justice initiatives under one roof; Fig. 8.2 shows their beautiful logo and organizational principles.

Lastly, land tenure is a very challenging issue for small farmers. Many are growing on land that the City (or some philanthropist) has allowed them to use. However, as we saw in Atlanta and New Orleans, it is quite common for cities (or land owners) to change their priorities and pull the

Figure 8.2. Cooperation Jackson (Mississippi): (left) Logo (right) Organizing Principles (Photos: David Schwartzman)

land right from underneath the farmers, be they individual or organizational operators. There needs to be stability built into the initial contractual relationships. Speculative investors sit on large urban properties waiting for the right time to sell. As a result, large areas of urban landscapes can remain fenced in asphalt rectangles for decades. Urban farmers, who are often reclaiming/using land that has otherwise been abandoned/ignored, deserve the utmost respect for the regenerative work they do, often in populated urban centers. Ordinances need to be passed that establish agricultural/ecological proposals as valuable enterprises, ones that should certainly be favored over those that wish merely to swing a profit at the expense of the neighborhood residents' quality of life and their ability to utilize available land.

8.4. Cuba's food path

Cuba's movement towards agroecology appears to have been launched in response to the dramatic reduction in oil supplies that followed the collapse of the Soviet Union. From 1972–1991 (referred to as the Soviet Period in Cuban history), the Soviet Union was a major exporter of oil and other resources to Cuba, while the United States embargo remained in effect (which commenced in 1958). During this period, Cuban universities and agriculture centers ramped up crop growth using an industrial model of agriculture, one based on heavy machinery and chemicals (pesticides and synthetic fertilizer) similar to what was being enacted in the USSR (Fernando Funes, personal communication, June 2017). Upon the dissolution of the Soviet Union in 1991, these imports declined substantially. In 1991, 90% of Cuba's petroleum came from the USSR (Boudreaux, 1991), afterwards, imports shifted to other nations and quantities were often less than 50% of 1991 levels (Quinn, 2006). Without the fuel to run the machines or create the chemicals (recall that pesticides and synthetic fertilizers are petroleum-based), Cuba had to shift methods and quickly. Agroecology to the rescue.

The 'Special Period' in Cuba, which followed the Soviet Period, witnessed a radical shift in agricultural methods. Without fuel to run machinery, human labor and hand tools became paramount. Compost (often produced through vermiculture techniques, i.e. using earthworms) replaced nitrogen-based synthetic fertilizer. And perhaps most significant was the establishment of urban farms on unused land as well as on rooftops, courtyards and patios (Quinn, 2006). While substantial in scope, the shift was insufficient to provide the same bounty, at least early on, and the average Cuban lost 20–30 pounds (Quinn, 2006; Cope, 2016) with per capita caloric intake dropping from 2,908 Calories to 1,863 Calories from 1989 to 1994 (Oxfam, 2001).

Oxfam's 'Cuba: Going Against the Grain' (2001) analyzes the policies and practices, beginning in 1992, that averted a national disaster. Many land reforms began soon after the 1959 revolution and additional ones following the Soviet Bloc collapse avoided the misery endured by '*campesinos* in so many [other] Latin American countries' (Oxfam, 2001). There were three key components to the reforms that kicked in soon after foreign agricultural implements and food became scarce. First, Cuba decentralized much of its farming operations. The introduction of a hybrid model, referred to as the Basic Unit of Cooperative Production (UBPC), where many farmers work through privately owned cooperatives (though the state remains the owner of the land) with other private farmers via usufruct ('rent-free lease agreements') (Oxfam, 2001). These cooperatives have a lot of control of what they grow and how they operate, and the farmers' earnings are connected to their production, encouraging them to be very industrious. Second, greater production incentives were provided by the state and more markets were made available to farmers, including farmers' markets and privately-run restaurants. Relatedly, large areas of Cuba shifted from sugar production, when prices plummeted, to vegetables (Oxfam, 2008). This agricultural diversification, which was partially made possible by the addition of irrigation systems and the construction of greenhouses, enabled household incomes to expand greatly while also encouraging the participation of women in agricultural activities as well (Oxfam, 2008). And third, Cuba introduced agroecological methods of producing food — noteworthy, 'organic fertilizer, animal traction, mixed cropping, and biological pest controls' became the norm (Oxfam, 2001). These methods were applied not only in the rural areas but significantly in quickly expanding urban settings as well. To give an idea of how impressive this activity was, consider that in 2001, approximately half of all vegetables consumed in Havana (largest city in Cuba with approximately two million inhabitants) were produced in urban gardens (Oxfam, 2001).

Ultimately, since these reforms were taking place within a socialist nation, Cuba enabled farmers to succeed by providing them with: (a) research centers; (b) organic pesticides; (c) (re-introduction to) animal traction; (d) new tools; (e) crop rotation and co-planting; and, (f) new irrigation techniques (Oxfam, 2001). And, these efforts fostered a return of Cuban diets to pre-1990 caloric levels despite the United States' very successful campaign keeping Cuba 'from receiving loans from the international finance institutions and [excluding] them from trade agreements' (Oxfam, 2001).

A summer 2017 educational tour organized by Food First revealed that Cuba's shift to agroecology continues with vigor. Finca Marta, a 20-acre farm outside of Havana, started only in 2011, now grows 60+ different varieties of vegetables, fruits and herbs, all organically and with vermicompost and manure. The farm also raises cattle, for meat and manure, and has many beehives, yielding 1.5 tons of honey in 2014 (Miroff, 2015) (see Fig. 8.3). Finca Marta employs terraced beds of vegetables and herbs to greatly reduce erosion and to combat/confuse pests,

Figure 8.3. Finca (Farm) Marta outside of Havana: (left) Apiary (for honey and pollinators) (center) Biogas operation using feces from farm animals (gas is used to cook the food in the farm's kitchen) (top right) Various seedlings, grown under netting (bottom right) Solar panels on top of livestock quarters (to pump water from well) (Photos: Carrington Morris, Peter Schwartzman & David Schwartzman)

Figure 8.4. Finca (Farm) Marta outside of Havana: Multicrop terrace farming, notice netting (used during summer) (Photos: Carrington Morris)

who find it very difficult to hone in on a particular crop with so many to choose from. Large permeable sheets cover much of the terraced landscape during the sunniest months of the year, enabling a bigger harvest because the plants are not baked as easily by the overly intense direct sunlight (see Fig. 8.4). Yet as technologically-sound as Finca Marta is, Fernando Funes Monzote, the farm's founder and head farmer, notes how important social aspects of the farm are to its productivity. Every work day, the farmers gather for a common lunch that includes tasty dishes made from the fruits and vegetables that they grow in the fields, cooked with biofuel generated onsite from manure (see Fig. 8.3 and Fig. 8.5). El Paraiso, another agroecological farm located near Viñales (west of Havana), integrates terraced agriculture with an onsite restaurant as well as a place for visitors to stay (see Fig. 8.6). Cuba has recently allowed private ownership of some of these businesses. Many such restaurants, called 'paladares', have started up. Some of these are focused on organic production and supporting agroecological farms; Fig. 8.7 shows one that even grows some of its own vegetables on its roof as well as supporting bee colonies on-site.

Another large agroecological farm, Organopónico Vivero Alamar (OVA), a Unidad Básica de Producción Cooperativa (Basic Unit of Cooperative Production) was established in a residential suburb of Havana; a one-hour documentary called 'Tierralismo' (Anderson,

Figure 8.5. Finca (Farm) Marta: A sample lunch full of a variety of vegetables. These daily meals bring the workers together to discuss experiences and strategies. (Photo: Carrington Morris)

Figure 8.6. Agrecological Farm (Finca Agrecologica) El Paraiso near Vinales, Cuba: Multicropping, companion planting, pollination planting, and restaurant and lodging onsite. (Photos: Carrington Morris)

Figure 8.7. El Jardin de Los Milagros, privately-owned restaurant focusing on organic food and food production: (left) Restaurant sign and roof top garden (background) where customers eat and the restaurant grows food (right) Rooftop scene showing plants growing in basins (below) and Beehives in wood boxes (for pollination) (Photos: David Schwartzman)

2013) — by film producer Alejandro Ramirez Anderson — presents its remarkable efforts. This cooperatively operated farm produces a multitude of vegetables, fruits and herbs on its 27 acres. Seedlings and ornamental and medicinal plants are also grown, often initially under a netting to keep pests out (see Fig. 8.8). OVA has well over 100 members and more than 20 employees. OVA sells its produce and other materials (such as compost and value-added products) to restaurants and directly to the public at its on-site shop. The proximity of the farm to a large apartment complex brings fresh food nearly to the front door of thousands of residents.

One of the more vivid observations of our agroecological tour of Cuba has to do with the Cuban diet. Surprising to us, Cubans eat a lot of meat (chicken and pork mostly), white rice and beans. Much of this rice must be imported and the production of this meat in Cuba relies on the import of GMO soy from Brazil to feed the animals. While the agroecological farms that we visited were capable of growing nearly any vegetable or fruit (as well as coffee), we heard again and again from the farmers that it is difficult to get Cubans to eat vegetables, which are known to contain high levels of important nutrients. The deep disconnect between what can be grown on Cuban soil, agroecologically and in great abundance, and that

Figure 8.8. Organopónico Vivero Alamar (OVA), cooperative farm, outside of Havana, Cuba: Growing vegetables under nets to control pests. (Photo: Carrington Morris)

which Cubans choose to eat demands much more attention. There exists cultural resistance in Cuba to changing traditional diets of rice/beans/pork to ones with more vegetables. This disconnect is observed in urban areas of the United States as well, where high-caloric, sugary (high glycemic) foods dominant the landscape of most grocery and convenience stores (due to their higher profit margins, long shelf lives, and calorie-demanding populace) while fresh fruits and vegetables are often hard to find (some areas are even designated food 'deserts' on this basis).

Even if we are able to grow adequate amounts of nutritious food with agroecological methods, this does not guarantee that people will choose to eat it. And if people do not eat it, farmers may feel compelled to shift to less nutritious options, a loss for society and the planet.

The shifting diet of the world's population towards more meat and calories (Oxfam, 2012) must be addressed as an issue on a grand scale, given the greatly increased negative impacts of agricultural methods tied to industrial meat production operations and grains grown to feed these animals. Tilman *et al.* (2002) note that consumers will need to have incentives to choose less environmentally-impactful food options. Many economic mechanisms exist to produce a shift in food purchasing; in the U.S. these include Farmers' Market coupons, subsidies for improved school lunches (which help establish lifelong eating habits), and government subsidies to fruit and vegetables farmers (which are currently almost non-existent). Considering how effective modern purveyors of processed food are in establishing dietary behaviors (case in point, the explosion of fast food restaurants worldwide), similar tactics could clearly be utilized to improve food consumption habits immensely. Given the ever-expanding understanding of incredible foods such as acai, moringa (also known as drumstick tree, horseradish tree, ben oil tree, or benzo tree), aronia (or red chokeberry), mangosteen, etc., education about the healthful benefits of eating lower down the trophic scale should help as well; Fig. 8.9 shows moringa, a leguminous plant which fixes nitrogen and can serve us nutritionally. The costs savings of

Figure 8.9. Moringa pods, leaves, and flowers. All are edible and highly nutritious. (Photos: Carrington Morris)

improving diets, in terms of reducing chronic health care expenses, are well-recognized (Schwenke, 2016; Sotos-Prieto *et al.*, 2017).

Enhancing educational efforts around food preservation is one significant response to misguided dietary choices and seasonal inconsistencies in produce availability. In Cuba, Vilda Figueroa and José Lama recognized this over two decades ago and their organization Community Food-Preservation Project (CFPP; Proyecto Communitario Conservacion Alimentos) launched in 1996 in the outskirts of Havana. In its twenty-two year history, it has produced dozens of publications (see Fig. 8.10), including those designed for children, who arguably should be the main focus for a dietary shift to more vegetables, to promote this effort (Cope, 2016). Their work, which received a jump start initially with the help of the Cuban government and which now works closely with parties involved with the slow food movement

Figure 8.10. Sample of publications produced by Community Food-Preservation Project (Cuba) focused on preservation & preparation food and cultural awareness. (Photo: Carrington Morris)

(particularly, the organization Slow Food), has taken off and become a fabulous resource for people all over the world (Cope, 2016). Though Cuba is in a subtropical climatological zone, which allows for year-round growing, food preservation provides extra food security by enabling people to can, bag (through drying), and ferment foods to maintain their important nutrients for later, often off-season, consumption. Many diets from wealthy regions of the world which ignore the seasonal growing cycles of their regions import 'out-of-season' fruits and vegetables from distant locations (often in different hemispheres), resulting in excessive CO_2 (and other GHG) emissions. The consumers of these diets could benefit greatly from preservation education and behavior. But, CFPP does not stop at preservation education, it also helps promote the growing (and eating) of highly nutritious (and climate-adapted) crops like cassava, by teaching people how to prepare it with a multitude of recipes and culinary techniques (Cope, 2016).

Cuba's ability to shift to a largely non-petroleum based agriculture provides an excellent example of what path may await the rest of humanity in the near future. However, Cuba's inability to reduce imports of food below 70–80% of total consumption indicates that there is still great work to be done (WFP, 2017). In the early 1990s, many staples for Cubans were imported at a very high clip — 90% of beans, 100% of wheat, and 50% of rice (Rosset and Benjamin, 1994). More recently, Cuba, despite the sizeable embargo still in place, still imports large amounts of wheat, corn, powdered milk, flour, soybean oil, and (still) ~60% of its rice (AGR, 2012). We found in our June 2017 tour, that the cutting-edge specialists in Cuba are very conscious of the potential shift to a Chinese/Vietnamese capitalist/commodity centered path of development once the blockade is terminated. Hence, there is the imperative of building a strong grassroots ecosocialist movement, especially among the youth.

As agroecological efforts in Detroit, Cuba and elsewhere continue, it is important to mention that Via Campesina, an international organization founded in 1993, has done tremendous work to move family-farm efforts forward and support food sovereignty efforts around the world. *Fertile Ground: Scaling Agroecology from the Ground Up* (Brescia, 2017) documents the progress of agroecological efforts throughout the global South and as such makes a critically important contribution to the scholarship in this burgeoning field.

8.5. The energy revolution

Despite what many people think, wind and solar energy installations have been growing at a feverish pace for the past twenty years. Despite a major global recession starting in 2008, renewable energy growth continues unabated. From 2000–2016, globally, solar PV energy capacity installation grew (on average) 40.6% annually and, from 1997–2016, wind power capacity installed grew (on average) 24.4% annually (calculated from annual data provided by WEC, 2016a; 2016b). These are phenomenal rates of growth particularly given the economic circumstances of the period. By the end of 2016, 308 GW of solar PV and CSP power capacity, and 487 GW of wind power capacity (Sawin *et al.*, 2017) had been installed; solar hot water capacity adds another 456 GW to the total energy capacity. Another fascinating fact, from 2012–2015, monetary investment in non-hydro renewable energy more than doubled that for fossil fuels and was about ten times more than nuclear power (Frankfurt, 2016; Geuss, 2017). In 2016 alone, US$242 billion was invested in renewable energy and fuels (Sawin *et al.*, 2017). It certainly appears that those with money to invest recognize that the energy of our future is renewable and they are cashing in during the early stages of the transition.

While the renewable energy (RE) growth rate is spectacular and mysteriously underreported by the mainstream press, the bigger story is that current rates of RE growth are likely sufficient to transition us to a full solar energy system quickly enough to avert much of C3, assuming they continue for the next 20–25 years. We conservatively calculate that to get to a 100% renewable energy system (providing some nine billion people in 2040 with 3.5 kW/person of power, reaching 41.7 TW by 2035, thus exceeding the estimated power globally necessary as discussed in Chapter 4) requires only 19 more years of the growth rates of the past 16–19 years (noted in previous paragraph); see Fig. 8.11 for a hypothetical look at the scale of the annual solar and wind installation required.[8.1] However, the

[8.1] These were calculated using a conservative 17% and 25% capacity factor for solar and wind, respectively. In 19 years, total energy production would be 365,000 TWh with 297,000 TWh (81%) from solar (PV) and 68,000 TWh from wind. This total amount is about three times more energy than produced globally in 2013. This calculation is very

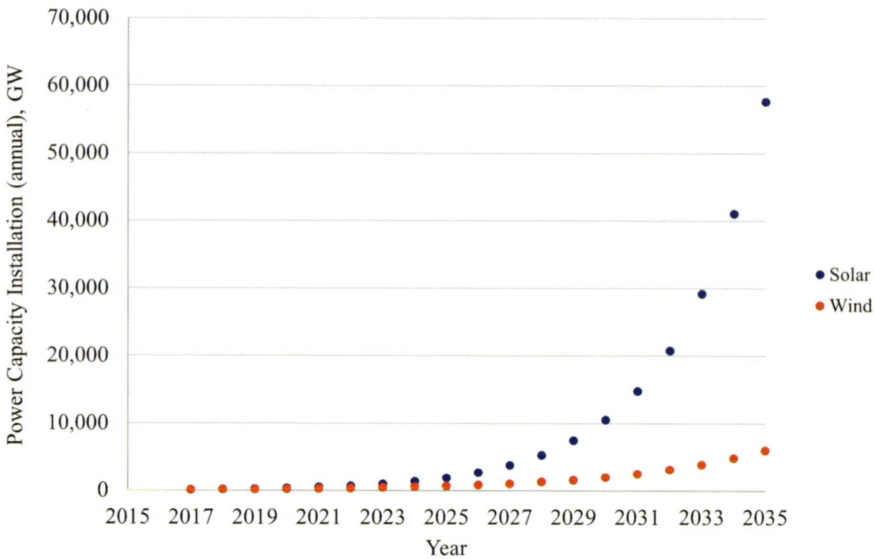

Figure 8.11. Annual wind & solar power capacity installation required to get to 100% wind/solar energy by 2035

Note: New installations begin in 2017 with 308 GW (solar PV) and 487 GW (wind) already in place. Annual growth rates are 40.6% (solar) and 24.4% wind. At the end of 19 years, 200 TW of solar and 31 TW of wind will be installed. Assuming, conservatively, capacity factors of 0.17 (solar) and 0.25 (wind), this will provide 41.7 TW. Recall, as discussed in Chapter 4, this effort will begin with fossil fuel investments but will increasingly be fueled by the installed wind and solar.

ability to continue to grow at these rapid rates into the future will require 'all hands on board' and rapid phasing out of fossil fuels (which become increasingly superfluous with time). Not only must we contend with multinationals who will not want to shrink their profitable investments in fossil-fuels, but large material resources will have to be expended to scale solar and wind to these levels (as discussed in detail in Chapter 4). Most importantly, the RE growth that has occurred to this point has been financed with massive capital by economic forces that are applying most of the same capitalistic tools that shaped our current food and energy

conservative as it considers no energy efficiency improvements or technological advances during the period of growth.

system. For this reason, we need to be very clear about how the Energy Revolution gets implemented. If we are not thoughtful about the way in which the solar energy system is created, global inequality will likely persist despite sufficient energy sources being available, analogous to the way we currently have enough calories to feed everyone but so many remain hungry or dangerously overfed. The Energy Justice Movement, as represented by The Energy Justice Network (Energy Justice, 2017) in the U.S. and the European Energy Justice Network (Energy Justice EU, 2017) in the European Union, are doing great work to bring these issues to the forefront of any energy transition. The early stages of Energy Justice as an academic discipline are covered by Jenkins *et al.*, (2016). Here, we take a look around the world to see how efforts to move solar and wind energy forward are being expressed.

8.5.1. *Cuba's energy path*

Cuba is planning a major push for renewable energy in the coming years. At a renewable energy conference in Havana in May 2015 (Ciercuba, 2017), Raúl García Barreiro, the vice minister of the Ministry of Energy and Mines, Cuba, made it clear that getting off foreign oil is a high priority for the nation. In particular, the Cuban government envisioned a chain of wind farms along the island's north shore, numerous 'bioelectric' stations using everything from sugar cane leftovers to pig manure, and solar installations of every size. Ferris (2015) gives us further insight into Cuba's process of solarization:

> "'Today, only 4% of electricity comes from renewable sources," Barreiro said, but he laid out plans for a drastic ramping up of efforts. In 2012, he stated, the Cuban government started crafting renewable energy goals to carry it through the period from 2014 to 2030. In August of 2017, it will begin to issue legal standards to support them. The island plans to get to over 20% utilization of renewable energy by 2020, and by that time to get 14% of its grid power from biomass, 6% from wind, 3% from solar and 1% from hydropower. Meanwhile, Cuba seeks to reduce the cost of delivered electricity from US$21.10 a kilowatt in 2013 to US$17.90 a kilowatt in 2020, and reduce the grams of carbon dioxide per kilowatt

from 1,127 in 2013 to 1,018 in 2020. There is evidence that Cuba's use of renewable energy is climbing steeply, said Daniel Stolik Novygrod, a photovoltaic specialist at the University of Havana. He said that Cuba's production of renewable energy grew from 3.2 MW in 2012 to 30 MW in 2014, and may reach 700 MW by 2018. The island wants 13 wind parks along the long north shore to produce 633 MW of wind, 19 bioelectric stations to produce 755 MW of "bioelectric" power and 700 MW of solar, Barreiro said. Cuba also plans to expand its hydropower capabilities, despite having short rivers that leave few sites untapped. Presently, the island has 142 hydroelectric dams, 32 of which are connected to the grid. The government recently identified 74 new sites that could together produce 56 MW of electricity, Barreiro said.'

8.5.2. *Can Venezuela lead a solar transition in Latin America?*

The only governments in the world to explicitly call their path ecosocialist are those of Venezuela and Bolivia (Ecosocialist Horizons, 2016), although Cuba's record especially with regard to initiatives in agroecology speaks to her lead in this respect.

As a major oil producer, Venezuela has the potential to significantly contribute to a solar energy transition, using the fossil fuel with the lowest GHG emission/energy consumed as an energy source to replace itself. In 2015, our senior author articulated this vision,

> 'We humbly propose to the people of Venezuela the outline of a plan by which Venezuela could lead a wind/solar power transition in Latin America using a small fraction of her liquid petroleum reserves, while still gaining revenue from oil exports as well as contributing to the same energy transition globally. Implementing this approach would be a critical component of Venezuela's self-identified path of ecosocialist development' (Schwartzman and Saul, 2015).

Here we emphasize our support for 'Oil for no one', referring to the heavy oil in the Orinoco basin, not the utilization of conventional oil reserves. The total proven reserves consist of mainly extra heavy crude (tar sands) of Orinoco, and if this is included, the reserves range up to 1000 billion barrels. Venezuela is already extracting heavy oil, and even supplementing the refined product with imported light oil. The provn reserves of conventional light to heavy oil in Venezuela are estimated to be 39 billion barrels, (excluding 259 billion barrels of extra heavy oil in the Orinoco basin; IESA, 2016, p. 21; Wikipedia, 2017c), although the further expansion of this reserve has been neglected in recent years (Ulmer and Parraga, 2014), particularly since the downturn in the economy following the sharp fall in the price of oil. With these available conventional reserves, we estimate that the goal of providing 3.5 kW/person can be achieved for the entire 400 million Mercosur population (comprising the South American trade bloc)—using our solar calculator, available at www.solarutopia.org. For an assumed EROEI ratio of wind/solar power equal to 25, this goal can be achieved in 15 years or less using 0.15 billion barrels of this oil per year (10% of the renewable is reinvested every year to create more of itself). Even this conventional oil will not be exhausted for the wind/solar energy transition in South America, lasting more than 30 years at a 1 billion barrel/year production rate (Schwartzman and Saul, 2015; Schwartzman, 2017).

Given the present situation in Venezuela and Latin America (these words written on August 26, 2017), we anticipate that what is proposed here for Venezuela's lead in solarization in Latin America will not be taken up until there is a turn to the left in this continent. However, judging from the personal experience communicated by Quincy Saul (the co-author of Schwartzman and Saul, 2015) who presented this proposal at meetings in Venezuela, those in support of the Bolivarian Revolution are very receptive to this proposal (Quincy Saul, personal communication, November 3, 2017). The senior author found the same response in his participation in the First Ecosocialist International Convocation in Venezuela, October/November 2017 (Schwartzman, 2017).

8.5.3. *Germany's Energiewende*

Hermann Scheer (2005; 2007), a Social Democrat member of the Bundestag, initiated Germany's remarkable solar/wind growth in recent years, the Energiewende (Energy Revolution). Feed-in tariffs were key to the acceleration in investment in renewable energy technologies, by providing producers with governmental funding above the retail or wholesale rates of electricity, thereby guaranteeing long-term security to renewable energy producers (Wikipedia, 2017a). On April 30, 2017, Germany got 85% of its electricity from renewables, a record (Hanley, 2017). In 2016, renewable energy supplied about 29% of power generation, and 12.6% of primary energy consumption, with the rest being derived mostly from fossil fuels (Appunn *et al.*, 2017). So, despite the hype about the Energiewende, Germany is still far from replacing fossil fuels and nuclear power with renewable energy. Nevertheless, the Energiewende has been coupled with a dramatic growth in community-owned and managed energy cooperatives — 800 in total — with 90% producing solar energy as of September 2016. Dr Andreas Wieg, of the Deutsche Genossenschafts-und Raiffeisenverband e.V, the federation for German co-operatives, shared his experiences with delegates at the 2016 Community Energy Conference:

> 'What we have learnt — the central message we share all over the world for community energy — is why it is important to have community, or citizen, energy projects…Without community involvement acceptance of renewable energy is harder. It is the sense of ownership that makes them work' (Wieg quoted in Harvey, 2016).

8.5.4. *Power to the People*

'Power to the People' is the title of an outstanding report of the Trade Unions for Energy Democracy (Sweeney *et al.*, 2005). This report comprehensively discusses the multiple ways that democratic control of electricity can be implemented, given a strong enough grassroots movement. In the U.S., the solar coop movement was pioneered in California (Weinrub, 2010). The Institute of Local Self-Reliance has also provided a very useful guide to how communities can gain ownership of renewable power (Farrell, 2016). And by no coincidence, the explicitly socialist online Jacobin magazine has featured an overview on socializing energy in the U.S. (see Aronoff, 2016).[8.2]

[8.2] Jacobin (2017) is a leading voice of the American left, offering socialist perspectives on politics, economics, and culture. The print magazine is released quarterly and reaches over 30,000 subscribers, in addition to a web audience of a 1,000,000 a month.

Aronoff (2016) argues for the following models for socializing energy:

(1) Municipalization, 'where city governments effectively buy out private utilities and run them in the public interest. This allows progressive city officials to enact ambitious plans for scaling back fossil fuels, and gives residents an accountable outlet for demanding a more dramatic transition. The most famous attempt at this has been in Boulder, Colorado, where residents voted in 2011 for the city to buy out monopoly provider Xcel Energy and create a model utility, aiming to get at least 54% of its power from renewables and offer lower rates over a twenty-year timeframe. Municipalization also offers a means of transition independent of austerity-stricken and spending-averse national governments. In Spain, Podemos-affiliated mayor Ada Colau announced a plan to municipalize Barcelona's power supply and shift toward 100% renewables usage starting in 2016.'
(2) Community Choice Aggregation (CCA) which allow 'city or county residents to "aggregate" their buying power through their local government and decide together where to get their electricity.'
(3) Following the precedent of the New Deal Rural Electrification Administration, establishing Energy Cooperatives: 'Today there are 900, serving 42 million people — 12% of the country's electric consumers — in some of its lowest-income and most conservative parts,' though most of the power now comes from coal.

These approaches are also vigorously promoted by the Next System Project, discussed in Chapter 9.

8.5.5. *China, contradictory trends abound*

Following up the brief remarks in Chapter 4 (What about the BRICs?) regarding China's potential role as a world leader in promoting a global ecosocialist transition, we revisit this possibility in light of the announced withdrawal from the Paris Agreement by the Trump Administration. As a result, China and India are now recognized as world leaders confronting climate change (New York Times, 2017).

Contradictory developments in China's energy/climate and food policies do indeed abound. On one hand China has ambitious plans to expand its nuclear power capacity (Cao *et al.*, 2016; Wikipedia, 2017b), on the other, she is the world leader in investing in wind/solar technologies (Bradsher, 2017; Forsythe, 2017) with a commitment to peaking coal use (Qi et al., 2016) in the context of a a 86% renewable energy goal by 2050 (Yang *et al.*, 2016). But simultaneously, China is investing heavily in mining in the global South:

> 'Over half of all public-sector lending from China to Latin America, some US$17.2 billion in 2017, went to the fossil-fuel industry…China is worsening the climate crisis with its financing elsewhere as well. From 2000 to 2015 China extended US$94.4 billion in loans to Africa [(CARI, 2017)], fueling extractive industries like oil, minerals and timber; the expansion roads and ports to get those raw materials to market; and dirty energy like large dams and power plants. Beijing is building and financing some 50 new coal plants across Africa [Rosen, 2017]' (Garzon and Salazar-Lopez, 2017).

The same authors conclude:

> 'China should approach its international projects with the same concern for the environment that it's starting to show at home. Beijing should refrain from supporting extraction in areas of global ecological importance, and instead heavily invest in clean, renewable energy projects. Civil society groups should keep the pressure on, and developing-country governments should incorporate such guidelines into bilateral agreements and project contracts. Continuing to pursue fossil fuel development is a losing proposition in the face of low oil prices, growing competition from renewables, and the scientific imperative to leave 80% of known fossil fuel reserves in the ground to avoid a catastrophic two-degree Celsius rise in global temperatures. A true climate leader would invest in the preservation of areas of global ecological importance rather than destroy them' (Garzon and Salazar-Lopez, 2017).

Acknowledging the incisive critique of this approach (e.g., Tanuro, 2008; Carbon Trade Watch, 2015), China's implementation of a carbon trading market may potentially facilitate reductions in carbon emissions and air pollution (Buckley, 2017), but these reductions are likely to be more than cancelled out by her fossil fuel investments in the global South.

While China's reforestation program is world-famous (Shockman, 2016; Macias-Fauria, 2018), she is apparently committed to another highly problematic approach, investment in GMO agriculture (Hvistendahl, 2017).

What offers grounds for optimism in regard to China's future, and indeed for the world, is the ongoing vigorous resistance of Chinese workers and citizens to the present mode of economic growth (Griffiths, 2016; Haas, 2016; Li *et al.*, 2016) as well as dissent even from Chinese intellectuals living in China (Ma and Xu, 2017).

8.6. Integration of food and energy

Though integrated, holistic efforts are necessary in the long run, many of the current efforts still modify one component of the system at a time — food, for instance, rather than food and energy. This highlights that opportunities for fuller integration await us with all the benefits anticipated from synergisms yet unrealized. Nonetheless, many early generation efforts show signs of how such collaborations/connections might come to fruition.

Of course, the significant efforts in the U.S. still work within the existing economic systems of capitalism. International efforts exhibit different patterns. Cuba, a noteworthy exception, saw their urban food movement manifest as a result of significant reduction in resource imports from the collapsing USSR. Detroit, Michigan found a new approach after having seen the massive movement of jobs overseas as multinational corporations sought bigger profits for their investors, through cheaper labor and weaker environmental regulations found abroad.

Working within capitalistic structures makes current sustainable efforts very tenuous ones. Many large urban farms don't own their own land and many are moved (or literally destroyed) at the whim of profit-driven property owners and investors, particularly in capitalist countries.

Efforts to expand solar panel production in some U.S. states and communities have been met by constraints set to restrict the autonomy of home owners and municipalities — 'free' market capitalism apparently does not apply to them, only to the largest, most powerful, corporations. Also, huge subsidies to the fossil fuel and nuclear industry since the 1970s forestalled the development of renewable energy. Gigantic subsidies to grain commodities over the past several decades have also marginalized honest appraisals of the horrific economic and health problems stemming from industrialized agriculture. Since these subsidies depend on decisions made by Congress, significant reforms will be forthcoming only by struggles in electoral politics coupled with strong movements for change.

We do not know exactly how or when we will get to OWSP. Will the great Transition shake out over several generations or will a Revolution occur much more quickly? If it is the former case then it will likely be in a much more challenging context, i.e., climate hell, than for the latter case. Will it be a peaceful change or one filled with brutal resistance from those currently benefiting from dominant systems? Will it come by traditional political efforts or by large scale reformist, but radical, grassroots movements? The answers to these questions make for great party conversation but arguably no one knows yet, likely not until the movements propelling this Transition get much stronger than at present. Yet, not knowing how the Transition/Revolution will occur does not prevent us from making the case in this book that a sustainable human civilization (with over nine billion people) is possible, and outlining how it seems to be getting started. The work to be done in both areas (food and energy) is huge, and while energy trajectories suggest that it is moving in the right direction with respect to its rapid growth (albeit primarily under the direction of capitalist investors), food's current trajectory is less clear, as the multinationals control the seeds and/or chemicals and have often succeeded in convincing impoverished/indebted/vulnerable nations to follow their lead. Nevertheless, in the next chapter, the last in this book, we further explicate what we think are the necessary conditions for making possible this transition on a time scale that is required to avoid the consequences of climate catastrophe. Indeed, we submit that bringing this transition/revolution into being is itself a necessary condition to prevent C3, realizing the OWSP.

Chapter 9

The Path to the Other World that is Still Possible

9.1. Prefiguring the future in the present

The concept of prefiguration in existing capitalist societies is critical to the emergence of a Global Subject with the capacity to create the other world that is still possible (OWSP), ending the rule of capital on the planet. We will first explore this important concept, then see how the Global Subject necessary and sufficient to create the OWSP may yet emerge.

Ernst Bloch's writings, especially *The Principle of Hope,* has reignited the consideration of the role of concrete utopias being born in global capitalism as a prefigurative mode of struggle for a world without the rule of capital. We find the concept of prefiguration invoked by Joel Kovel (1992), who regards *The Principle of Hope* as 'absolutely core for ecosocialist discourse' (Kovel, 2017). Prefiguration captures the vision of 'eroding' capitalism, while reforms such as the nationalization of the energy industry are close to the category of 'taming' capitalism (Wright, 2015, 2016). Ana Dinerstein, a political sociologist at the University of Bath, frames the concept of prefiguration:

> 'In order to grasp the radicality of "concrete utopias", Marx's critique of political economy should be read through the lenses of Bloch's "principle of hope." The prefigurative critique of political economy enables us first to understand concrete utopias as struggles for alternative forms of "life" amidst the crisis of social reproduction. Second, the prefigurative

> critique of political economy enables us to address the problem of the subordination of concrete utopias to West/North/Eurocentric ways of conceiving of revolutionary change and to decolonise them' (Dinerstein, 2016a).

Further,

> 'the principle of hope (Bloch's masterpiece) was originally titled "Dreams of a better life"...I think that to produce these concrete dreams is the fundamental task of Marxism today, insofar as concrete utopias are the forms of organising the struggles for alternative forms of social reproduction towards a dignified and human life in South Asia, Latin America, Europe and the world' (Dinerstein, 2016b).

Post-capitalism will either mean a return to a pre-capitalism in the abyss of climate catastrophe or alternatively the creation of a much more hopeful future. This is the imminent bifurcation facing humanity. As we rapidly approach this great bifurcation, the vision of Blochian Marxism coupled with liberation theologies are critical in avoiding the catastrophic outcome. Two examples of prefiguration are considered here: expanding the commons and the Universal Declaration of Human Rights. To return to the Hegelian metaphor 'the negation of the negation,' a commonly cited example by Marxists, we begin with the first negation of primitive communism into class-stratified societies, and then pass to the second, the final negation into communism, from prehistory to history. We argue that the growth of movements around our globe to 'defend, extend, and deepen the commons' (Wall, 2005) is the beginning of the second negation within the womb of capitalism itself. The ancient commons of humanity, the oceans, forests, grazing lands, are now joined by 'open source' in cyberspace, a contested terrain in the struggle to decommodify information (Davis *et al.*, 1997; Dyer-Witheford, 1999; 2015). Social management of communally owned land is of course found even in capitalist countries in the form of national parks, biosphere reserves, and recognition in international law of the common heritage of humanity — such as Antarctica, the oceans, the atmosphere, and the Moon. This social management will be vastly expanded in an ecosocialist transition out of capitalism.

Besides waging defensive battles against austerity, social-service cuts, and the privatization of public resources, expanding the ecosocialist horizon is a process of 'eroding capitalism' (Wright, 2015; 2016), going on the offensive for social governance of production and consumption including the following objectives:

(1) Creation of decentralized solar power, food, energy and farming cooperatives and worker-owned factories: a 'solidarity economy' (Alperovitz, 2013; 2016; but see a critique of this approach pointing out its limits from Gindin, 2016).
(2) Publicly owned and accountable banks;
(3) Municipalization of electric and water supplies;
(4) Nationalization of the energy, rail, and telecommunications industries; and,
(5) Compulsory licensing of state-of-the science wind/solar and information technologies, making them freely available globally, following the precedent of the U.S. Clean Air Act.

The Universal Declaration of Human Rights is a prefiguration of potential alternative social rights and relations. The adoption of the Universal Declaration of Human Rights (UDHR) on December 10, 1948 by the General Assembly of the United Nations was inspired by the victory over fascism in the Second World War. The critical role of the Soviet Union in making this victory possible forced the inclusion of economic, social and cultural human rights language into this historic document, although Western governments succeeded in splitting off these rights in legally binding covenants starting in 1951 and have continued to resist implementation (Fields, 2003). Nevertheless, human rights discourse has emerged as a powerful asset in this struggle, even within the United States.

Life expectancy statistics demonstrate that the vision of the UDHR and conventions that followed remains unfulfilled in this world of continuing great disparities dominated by the reproduction of capital. As Bill Bowring (2008) puts it:

> '… neither the concept of human rights nor that of social justice can have content, meaning and significance except through their complete

> reinvention and reintegration in the real activity of women and men in the always turbulent and dangerous world into which they are thrown' (p.166).

A critical complement to universal human rights is the right of future generations to continue to enjoy the existing biodiversity of our planet, and even the partial restoration of what has been largely eliminated (e.g., prairies). The historic Convention on Biological Diversity entered into force on December 29, 1993. The convention has been ratified by 196 countries, but not by the United States, a fact that greatly weakens its implementation (Convention on Biological Diversity, 2017). The protection of biodiversity is closely tied to the imperative of replacing present unsustainable energy sources with wind and solar, the conversion of agricultural systems to agroecologies, and of course implementing an effective prevention program to avoid catastrophic climate change (C3). Biological research demonstrates that biodiversity plays a key role in ecological resilience (Peterson *et al.*, 1998; Thompson *et al.*, 2009).

But what is the theory powerful enough to confront these challenges? James O'Connor (1988), founding editor of the journal *Capitalism Nature Socialism*, argues:

> 'In ecological Marxist theory, the struggle over production conditions has redefined the class struggle beyond any self-recognition as such, at least until now. This means that the capitalist threats to the reproduction of production conditions are not only threats to profits and accumulation, but also to the viability of the social and "natural" environment as a means of life.'

Since these threats are global, so must be the struggle to confront them. Thus, only a transnational Subject will potentially have the power to create the OWSP. Recognizing this goal is not a deterministic outcome because we wish it into being, it is rather highly contingent on how local and national struggles converge with sufficient power on a timescale limited by the magnitude of the threats to the environment, most of all climate change.

9.2. Prefiguration of the transnational subject

Thus, the global class struggle to protect ecological sustainability must be a central objective of 21st Century ecosocialism. It is clear that the divisions in the global working class and potentially allied strata, that are derived from the exploitation of differences in nationality, ethnicity and race, gender, sexual preference, etc., must be significantly overcome if there is to be any hope of creating a countervailing force to national and transnational ruling classes in the global struggle for power. Of course, these divisions have a long history of utility to ruling classes. Overcoming them can ignite a force of unprecedented dimensions for ending the rule of capital. A transnational ecosocialist movement must emerge as a component for the success of this struggle — in the course of which traditional forms of organization will be superseded through the immense creativity of those involved in the global peace and climate justice movement, especially youth (Wagar, 1989; 1992; 1999; Schwartzman, 1992a; 1992b; Kovel, 2002; Dyer-Witheford, 2015). We see these forms emerging in embryo in the networking and inventions of solidarity in the World Social Forum and Occupy movements, and in particular, in the women's and indigenous peoples' struggle. Transnational labor at the core of global peace and justice movements must by necessity emerge as the central countervailing force to transnational capital, which increasingly contests nation-centric forms of capitalist accumulation (Robinson, 2004; Harris, 2005).

The World Social Forum process is the embryonic crystallization of a World Party, the name given by W. Warren Wagar (1989; 2001) to a transnational political force representing the common interests of humanity on all spatial scales, including in national and local struggles. If an ecosocialist transition is the goal, then first the power of capital must be constrained, then finally removed and replaced by social management of both the physical and political economy.

But, can the defects in capitalism we have outlined in previous chapters be corrected, short of moving to a post-capitalist society, by a process of reform? The reformability of real existing capitalism must be tested by actual class struggle, in other words by defending and expanding democracy in the social, political and economic spheres.

It is precisely now, as global capitalism faces the converging economic, social and climate crises, that struggles for a Global Green New Deal (GGND) should emerge as the focus for an ecosocialist movement.

Thus, the critical relationship of the global North and South must be a focus of transnational class struggle. As Battistoni (2016) illuminates in a must-read examination of Rosa Luxemburg's legacy:

> 'One of Luxemburg's major contributions in The Accumulation of Capital was to point out that capitalism is dependent on an "outside" — that is, on having non-capitalist societies and forms to draw on as it continually expands. The capitalist economy needs an external source from which to obtain resources, find cheap labor, sell surplus commodities, and so on. Capitalism, she writes, "needs non-capitalist social strata as a market for its surplus value, as a source of supply for its means of production and as a reservoir of labor power for its wage system. Yet this is a self-defeating process: once capitalism becomes the only form of production, it can no longer function because it has no outside to draw on"' (p. 70).

9.3. Looking beyond capitalism: the Next Systems Project

In 2015, a call was issued by a broad group of scholars and activists to form the Next Systems Project (2017). While it focuses on the United States, this project has implications for the entire global community. Here is a summary of its purpose:

> 'The Next System Project is an ambitious multi-year initiative aimed at thinking boldly about what is required to deal with the systemic challenges the United States faces now and in coming decades. Responding to real hunger… and building on innovative thinking and practical experience with new economic institutions and approaches being developed in communities across the country and around the world, the goal is to put the central idea of system change, and that there can be a "next system," on the map.
>
> Working with a broad group of researchers, theorists and activists, we seek to launch a national debate on the nature of "the next system" using the best research, understanding and strategic thinking, on the one hand,

and on-the-ground organizing and development experience, on the other, to refine and publicize comprehensive alternative political-economic system models that are different in fundamental ways from the failed systems of the past and capable of delivering superior social, economic and ecological outcomes.

By defining issues systemically, we believe we can begin to move the political conversation beyond current limits with the aim of catalyzing a substantive debate about the need for a radically different system and how we might go about its construction. Despite the scale of the difficulties, a cautious and paradoxical optimism is warranted. There are real alternatives. Arising from the unforgiving logic of dead ends, the steadily building array of promising new proposals and alternative institutions and experiments, together with an explosion of ideas and new activism, offer a powerful basis for hope' (Next Systems Project, 2017).

Further, from its website:

'TIME TO FACE THE DEPTH OF THE SYSTEMIC CRISIS

There are political-economic system models that deliver superior social, economic and ecological outcomes.

<u>What's at Stake</u>

The challenging realities of growing inequality, political stalemate, and climate disruption prompt an important insight. When the old ways longer produce the outcomes we are looking for, something deeper is occurring. We are at or near the bottom among advanced democracies across a score of key indicators of national well-being — including relative poverty, inequality, education, social mobility, health, environment, militarization, democracy, and more. We have fundamental problems because of fundamental flaws in our economic and political system. The crisis now unfolding in so many ways across our country amounts to a systemic crisis.

Today's political economic system is not programmed to secure the well-being of people, place and planet. Instead, its priorities are corporate profits, the growth of GDP, and the projection of national power. Large-scale system change is needed but has until recently been constrained by a continuing lack of imagination concerning social, economic and political alternatives. There are alternatives that can lead to the systemic change we need.

What's next?

It is time to explore genuine alternatives and new models — "the next system." It is time to debate what it will take to move our country to a very different place, one where outcomes that are truly sustainable, equitable, and democratic are commonplace. Let's begin a real conversation — locally, nationally, and at all levels in between — on how to respond to the profound challenge of our time in history.

We need to think through and then build a new political economy that takes us beyond the current system that is failing all around us. Systemic problems require systemic solutions. We must think boldly about what is required to deal with the systemic difficulties facing the United States.

An extraordinary amount of experimentation is taking place in communities across the United States — and around the world. These sophisticated and thoughtful proposals for transformative change suggest that it is possible to build a new and better America Those of us signing this statement are committed to working towards these ends' (Next Systems Project, 2017).

9.4. Post-capitalism and the end of value

9.4.1. *Value production under capitalism*

In Marxist theory, labor power produces Value in the process of capital reproduction, the essence of the capitalist system (see a lucid exposition in Heinrich, 2004). But what about other sources of value? Does nature produce value in this economic cycle of capital reproduction, as some in sympathy with Marxism claim? For example, do bees produce value in this sense, since they produce honey which can be sold for a profit in the capitalist marketplace? This very question was the basis of an illuminating dialogue between a Marxist geographer and an ecological economist (Kallis and Swyngedouw, 2017). In this dialogue, the ecological economist Giorgos Kallis asks, 'What is the main Marxist argument against accepting the simple — in my view — fact that nature does work? …

Could we say then that resources "produce" value? If not, why not?' The Marxist geographer Erik Swyngedouw responds:

> 'Honey, as the work of bees, has both use-value and exchange-value, but no Value. Capitalists are not interested in use-value per se; they are interested in surplus value. Labour power, also a commodity, produces more Value during a given period than the value of labour power itself. This difference is surplus value. This reasoning holds only for capitalism. I do not produce Value when I work in my garden; a hunter-gatherer in Amazonia does not produce Value. Labour is the capacity to work and we all have that. But only labour power has Value, as the socially necessary labour time to reproduce labour, which is bought by the capitalists for a certain time. The capitalists' trick is that labour power produces more value than the labour required to reproduce. No surplus value, no profit and no capitalism. Bees, microbes, algae, rivers do all sorts of useful labour. For wild honey, the price that the one who takes it from the beehive gets is Rent (he gets a price for the honey because he owns it/appropriated it). The bees worked "for free"' (Kallis and Swyngedouw, 2017).

The argument that we often have about non-human labor is inspired by a moral injunction to recognize the great contribution that bees, etc., make to life. That is undisputed. But Marx was trying to explain exploitation, the production of surplus value, while maintaining the equivalence of exchange. Swyngedouw sums up by saying, 'It is not Marxist theory that has limits, it is the actual practice and workings of capitalism that have gigantic limits. And with respect to nature, it is precisely that it is not valued' (Kallis and Swyngedouw, 2017).

Huber (2017) fleshes out the same point:

> 'In this theoretical intervention, I argue that Karl Marx's theory of value remains a powerful way to understand nature–society relations under capitalism. I suggest environmentalist critiques often misunderstand Marx's value theory as a theory that "values" workers over nature. His critical theory is better understood as an explanation of how capitalist value exploits both workers and the environment. My defense of Marxian value theory is articulated through five "theses." I provide

> empirical illustration based on recent research into the nitrogen fertilizer industry. (1) Value theory does not refer to all values. (2) Marx's contention that nature does not contribute to value helps us explain its degradation under capitalism. (3) Marx's value theory rooted in production allows for a critique of environmental economic valuation schemes (e.g. payments for ecosystem services) which are based on neoclassical value theories rooted in consumption/exchange. (4) Value is abstract social labor, but that means it also abstracts from nature. (5) Capital does value certain parts of nature and that matters. I conclude by advocating a "value theory of nature" in the spirit of Diane Elson's powerful articulation of Marx's "value theory of labor"' (Abstract).

Lebowitz (2003) has argued that the conditions for the reproduction of labor power is itself a site of multi-dimensional class struggle, with profound insights into how this struggle necessarily entails the development of human potential, only achieving full fruition in self-managed socialism (Lebowitz, 2010). A summary of a speech Lebowitz made in 2017 analyzes the previous defects of 20th century socialism based on the limits of Marx's theory of capital:

> 'Unfortunately, for many who have followed Marx in name and others who never pretended to do so, there is only one product — the change in circumstances, the change in the object of labour. The second product — the change in human beings, the change in the subjects of labour — is ignored. The political effects of this blindness can be seen everywhere. In the countries of "real socialism" where the absence of self-government and self-management produced a working class with neither the capacity nor the will to prevent the restoration of capitalism. In the social democrats who, convinced that they are cleverer than capital, use the strength of the working class as a credible threat in their negotiations rather than as a force to be built and built and, accordingly, emerge from the most disgraceful defeats as immaculate as they were innocent. In political parties of the left which, rather than treating social movements as multiple sites for developing the capacities of the working class, view them as fertile ground for the recruitment of cadres for their disciplined phalanxes and celebrate in their solitary gatherings the distilled purity of their brands and their

> preparedness for the next October...It is not only political practice, however, that has suffered from the eclipse of the second product. Without an understanding of the centrality of the key link between human activity and human capacity, we are blind to the limitations of Marx's *Capital*' (Lebowitz, 2017).

David Laibman (2012) stresses the critical role of class struggle in determining this valorization: 'Labor power is, always and necessarily, a special commodity, never subject to full valorization like other commodities. Its value is always the outcome of the balance of class forces ("balance" here in the sense of "relationship" or "correlation," with no implication of "equilibrium" or any sort of inherent equality or consistency)'.

Class struggle is the mode of self-valorization of labor power by virtue of the activity of the working class itself. Dinerstein (2015) points to this concept explored by Harry Cleaver:

> '"Self-valorisation" designates "the ability of workers to define their interests and to struggle for them–to go beyond mere reaction to exploitation, or to self-defined leadership, and to take the offensive in ways that shape the class struggle and define the future" (Cleaver in Cleaver and De Angelis, 1993). Autonomist Marxism provided a general line of reasoning on working-class self-activity and the politics of a diversity of movements and ideas within the Marxist tradition (Cleaver, 2011, 54). The "inversion" of the class perspective advocated by Tronti (Cleaver, 1992) centres the analysis on class struggle rather than on capitalist development. Cleaver highlights that the term "Self-valorisation... focus[es] attention on the existence of autonomy in the self-development of workers vis-à-vis capital" (Cleaver and De Angelis, 1993) rather than as a derivative of the development of capital. Self-valorisation refocused attention onto the struggle against capital and for new forms of being: it involves "a process of valorization which is autonomous from capitalist valorization — a self-defining, self-determining process which goes from mere resistance to capitalist valorization to a positive project of self-constitution" (Cleaver, 1992, 129)' (pp. 38–39).

But this process of self-valorization within the capitalist society is not only creating prefigurations of the future, but is a critical component of the valorization of labor power, and hence advances the power of the working class breaking out of the constraints of capital reproduction towards the ecosocialist horizon.

Fanelli and Noonan (2017) emphasize the connection between ongoing class struggle in capitalist society with the realization of real freedom possible in post-capitalist socialist civilization:

> 'If the whole point of socialism is to replace a society in which need-satisfaction is subordinate to the accumulation of capital, to ensure that resources and social institutions enable the expression and enjoyment of human life-capacities in forms of activity that are meaningful to the agents and valuable to the lives of others over an open-ended human future, then struggles that free the lifetime of mortal individuals from alienated labour, even if they do not lead to the revolutionary overthrow of capitalism, cannot be regarded as irrelevant to that overall project, precisely because they accomplish to a limited extent that which the struggle for a socialist alternative to capitalism hopes to realize absolutely: the satisfaction of the social conditions for all round self-realizing freedom' (p. 147).
>
> 'What does the creation of universally accessible public institutions mean? The re-channeling of wealth away from private accumulation towards life-requirement satisfaction on the basis of *need,* not the ability to pay. In other words, the funding of universally accessible public institutions through taxation is another inroad against the power of capital over life. When education, health care, access to cultural institutions, and pensions are taken out of the cycle of commodified exchange and made available to all people *on the basis and to the extent of their needs for them*, real life improves…' (pp. 155–156).

Peter Frase (2015) conceptualizes class struggle in relation to the workplace, arguing that 'the strengthening of the working class both inside and outside the workplace becomes the force that pushes us toward the utopian ideal of a post-scarcity society and the abolition of wage labor'. Further, Frase (2016) elaborates on this theme, particularly in his chapter on 'Communism: Equality and Abundance', the most optimistic of the four futures he considered. Beginning as an outcome of class struggle under capitalism, the decommodification of labor is the foundation of his vision of communism. His examples include socialized medicine and guaranteed income protection in retirement and unemployment protection, culminating in a basic income independent of work, the result of 'non-reformist reform' struggles, appropriating a phrase from Gorz (1967).

Drawing from the analysis of Marx's 'Fragment on Machines,' Paul Mason (2015), too, argues that post-capitalism is inevitable because of the elimination of scarcity driven by the ever-growing power of the 'general intellect'. Information technology expands its reach into every aspect of life. 'The class struggle becomes the struggle to be human and educated during one's free time' (Mason, 2015, pp. 136–137).

Mason sees the ongoing increase in productivity growing out of both high efficiency renewable energy and information technologies as a serious potential basis for the system beyond capitalism. Jeremy Rifkin (2014) has also recently argued that high efficiency renewable energy and information technologies are posing an alternative beyond capitalism. But only global class struggle has the capacity to realize this potential for all of humanity and not simply a privileged elite living in gated communities.

However, Nick Dyer-Witheford (2015) finds the impact of the 'general intellect' under really-existing capitalism contradictory, with a 'moving contradiction' between:

> 'the encompassing of the global population by networked supply chains and agile production systems, making labour available to capital on a planetary scale, and on the other, as a drive towards the development of adept automata and algorithmic software that render such labour redundant' (p. 14).

However, it is important to note that several prominent Marxist scholars disagree with Marx's formulation in 'The Fragment on Machines', written before *Capital*. For example, Michael Heinrich (2013) argues that Marx confused immediate labor-time with abstract labor, which embodies the substance of value, and this confusion was gone in *Capital*.

Nevertheless, the discovery of 'The Fragment on Machines' in the 1960s and 1970s has, in the 21st Century, ignited a vision of the end of the rule of capital on our planet by utilizing the cutting edge of science and technology (Caffentzis, 2013).

9.4.2. *What is 'multi-dimensional class struggle'?*

Multi-dimensional class struggle is waged at every level, from the workplace and the classroom to the globe (transnational), and at every intersection of the oppressed and the exploited (race, gender, sexual orientation, ethnicity, citizenship status, religion, age, degree of able-bodiedness). But intersectionality should be unpacked with Marxist analysis, as done with great clarity by Victor Wallis (2015) who emphasizes that class domination is its cement or binding agent.

Theodore Allen's (1997) seminal book, *The Invention of the White Race,* should be mentioned. The book documents how, after the suppression of late 17th Century Bacon's Rebellion in colonial Virginia, indentured laborers of European descent were given 'white skin privileges' over those of African descent by the ruling class as a means of social control of the work force. Following Allen's analysis, this was the root of the system of white supremacy continuing to the present.[9.1]

In addition, the centrality of women's oppression must be recognized for its revolutionary potential (Federici, 2012; James, 2012; Salleh, 2014). The range of women's struggles in recent years is astonishing, from protection of the forest commons and of the Niger Delta, to the revolt against unwaged housework. The latter has directly confronted the valorization of labor power. This concept as well as other aspects of 'social reproduction theory' are now the subject of wide-ranging discussion (Bhattacharya, 2017).

9.4.3. *Solar communism and the end of value*

The senior author came up with the concept of 'solar communism', which refers to the culmination of a global ecosocialist transition from capitalism (Schwartzman, 1992a; 1996; 2016a). Solar *communism*, oh the notorious 'C' word! Before the reader draws the wrong conclusion, i.e., that this concept of communism is equivalent to a return to the worst aspects of the 'real existing socialism' of the 20th Century (countries led by parties

---

[9.1] Jeffrey Perry (2017) has a very useful website with slide presentations and other materials on this subject. Perry is the biographer of Hubert Harrison, a very important Black leader from the early 20th Century.

calling themselves communist), it is rather a return to Marx's own vision, with an update for this century. Solar communism is defined as a global civilization realizing Marx's aphoristic definition of communism (Marx, 1891) for the 21st Century: 'from each according to *her* ability, to each according to *her* needs,' with *her* referring to both humans and ecosystems. Marx's phrase 'From each according to his ability, to each according to his needs' actually has roots in the Bible, specifically in the New Testament:

> 'In Acts of the Apostles the lifestyle of the community of believers in Jerusalem is described as communal (without individual possession), and uses the phrase "*distribution was made unto every man according as he had need*"' (Wikipedia, n.d.c.).

Apparently, its invocation just prior to rise of the socialist movement was by Etienne-Gabriel Morelly:

> 'In his 1755 Code of Nature "Sacred and Fundamental Laws that would tear out the roots of vice and of all the evils of a society," including:

I. Nothing in society will belong to anyone, either as a personal possession or as capital goods, except the things for which the person has immediate use, for either his needs, his pleasures, or his daily work.
II. Every citizen will be a public man, sustained by, supported by, and occupied at the public expense.
III. Every citizen will make his particular contribution to the activities of the community according to his capacity, his talent and his age; it is on this basis that his duties will be determined, in conformity with the distributive laws' (Wikipedia, n.d.c.).

Morelly apparently influenced the communist of the French Revolution, Gracchus Babeuf, who in turn inspired the utopian socialists of the 19th Century, then Marx and Engels (Bertsch, n.d.; Birchall, 2016). An interesting discussion of the aphorism's modern application is found in Gilabert (2015), although this paper, unlike Marx (1938), attributes it to the socialist, rather than the communist principle.

Schwartzman (2016a) puts some historical context behind his invocation of solar communism:

> 'Communism was an inspiring vision to millions in the twentieth century, with well-known failures in its realization as "really existing socialism," or what some Marxists prefer to call "state capitalism". The construction of socialist societies in the twentieth century occurred in a very unfavorable context, under continuous attack by capitalist powers – starting with initial intervention soon after the Russian revolution, followed by the Second World War and the cold war. Socialist societies of the twentieth century had both real internal achievements and well-documented state-sanctioned mass suffering and death, in parallel with immense positive impacts on global politics, including the defeat of fascism, the post-Second World War end of colonial oppression, and the development of social-welfare programmes by capitalist states challenged by the benefits for working classes in socialist countries (e.g., West Germany and the German Democratic Republic)...Really-existing twentieth-century "socialism" (and its survivals into the twenty-first century) combined characteristics of communism, capitalism and state capitalism. But this is no surprise, given the impurity and complexity of a real transition from capitalism, potentially into communism...Jodi Dean [2012] reasserts the vision of radical materialist utopia that has been buried, reburied, yet never extinguished. But to her invocation of a communist horizon in the twenty-first century one must add that this will be solar communist. An ecosocialist horizon is imperative to prevent and prefigure activity embodied in multidimensional class struggle in our world dominated by the rule of capital; to prevent catastrophic climate change and along the way demilitarize, solarize, and transform agriculture with agroecologies; to prefigure the future in the present by expanding the commons, virtual and material; and to move toward the ecosocialist horizon, reaching it and moving toward the solar communist horizon' (pp. 145–146).

And regarding the 'solar' part of solar communism:

> 'Under "really existing capitalism," solar is also the energy source most compatible with decentralized, democratic management and control, relatively free of the dictates of the Military Industrial Complex (MIC), compared to

> fossil fuels and nuclear power. Finally, a solar "clean energy" transition is a critical component of the Global Green New Deal, and an ecosocialist path out of capitalism. But this vision has nothing in common with the stereotype of one-party dictatorships; if realized it will be the product of bottom-up struggles, a profoundly democratic process' (pp. 145–146).

So how does this relate to the end of Value, central to the reproduction of capital? This passage from Jim Davis (2000) is illuminating:

> 'New technologies express the fulfillment of Marx's writings in his "Fragment on Machines" — a production system without human labor, where the productivity of technology so overwhelms the production process that "labor time ceases to be the measure" of wealth and "production based upon exchange value collapses"... I argue that as a historical category, Value has at least a theoretical end...The new technological climate does not in itself destroy the Value system, or capitalism, but it does create the conditions for Capital's destruction and the construction of a communist society. The end of Value is not automatic, but a conscious act by class forces born out of the new conditions.'

Davis ends his essay saying, 'This is how Value will end — as a political act, the exercise of class power.' A critical material prerequisite of the end of value is the availability of virtually free abundant energy derived from a global infrastructure of high efficiency capture of solar energy. This infrastructure will create the supply and quality of energy necessary to radically reduce negative environmental impacts, indeed to also restore and repair both the technosphere and ecosphere, though irreversible damage has already occurred (e.g. biodiversity loss). In contrast, continued reliance on the present unsustainable energy supply not only contributes to well-known negative environmental, ecological and health impacts, but also thereby reduces labor productivity while externalizing the costs of these negative impacts. Meanwhile the capitalist state subsidizes fossil fuels and nuclear power while cutting budgets for health and environmental protection. As previously cited, the huge subsidies going to fossil fuels, estimated in a recent IMF study at US$5 trillion a year, also contain indirect costs including health impacts from air pollution which result in an estimate of 3–7 million deaths every year.

9.4.4. *Committed science and communities of struggle*

This is our moral imperative: every child born on Earth has the right to a full life of creative fulfillment, to an environment free of hatred and pollution, and to a world with what is left of our planet's biodiversity intact. This will require the termination of the present global regime prioritizing capital reproduction over human needs as well as those of nature.

In our political practice, it is essential to: first, recognize the potentiality of the moment and act, else we lose the chance to change the future; and, second, utilize every division in the ruling class to gain the necessary political momentum to prevent catastrophic climate change (C3). But the vision itself and its realization can only come about as a product informed by the dialogue between a committed scientific/technological intelligentsia and communities of struggle, as embryos of the future are created within the womb of globalized capitalism, and as global class struggle unfolds to achieve its full reality.

Socialist or Marxist political economy is necessary but not sufficient in itself to advance a vision of 21st Century socialism. This vision must fully engage the natural, physical and informational sciences — in particular, climatology, ecology, biogeochemistry, and thermodynamics — as well as take full account of the wisdom derived from the experience of thousands of years of indigenous peoples' agriculture and culture. This will inform the technologies of renewable energy, green production, and agroecologies, whose infrastructure are to replace the present unsustainable forms.

9.5. Conclusion

The great Italian Marxist Antonio Gramsci in his December 19, 1929 letter from prison (where he languished for 11 years) said, 'I'm a pessimist because of intelligence, but an optimist because of will' (Gramsci, 2011). Putting this advice into the present context, in an interview about his latest novel, *New York, 2140*, the science fiction writer Kim Stanley Robinson says:

> 'It's about climate change and sea level rise, but it's also about the way that our economic system doesn't allow us to afford a decent future. As one of the characters says early in the book, "We've got good tech,

> we've got a nice planet, but we're fucking it up by way of stupid laws." Finance, globalization — this current moment of capitalism — has a stranglehold on the world by way of all our treaties and laws, but it adds up to a multigenerational Ponzi scheme, an agreement on the part of everybody to screw the future generations for the sake of present profits. By the logic of our current system we have to mess up the Earth, and that is crazy. My new novel explores this problem and how we might get out of it' (Billings, 2017).

In his essay in *State of the World 2013*, Robinson looks forward to a post-capitalist world concluding:

> 'We can see our present danger, and we can also see our future potential: a stable human population of some seven to nine billion, living cleanly and well on a healthy biosphere, sharing Earth with the rest of the creatures who rely on it. This is not just a dream but a responsibility, a project. And things we can do now to start on this project are all around us, waiting to be taken up and lived' (Robinson, 2013, p. 380).

We take this project as our goal and our responsibility to our own children and grandchildren — indeed to all of our planet's children and grandchildren — who will inherit what we can collectively achieve, succeed or fail in preventing climate catastrophe. Use our book as a resource and act!

Epilogue

As we finish our examination of the final proofs, making minimal corrections, we offer these updates citing very recent and relevant publications.

Introduction

Following up on our discussion of Hurricane Harvey, we take note of the devastating impact of Hurricane Maria on Puerto Rico, whose people still bear the twin burdens of neocolonial and neoliberal policies; a Harvard University study estimates that over 4,600 people died (Kishore *et al.*, 2018), far more than the initial official report documents.

Chapter 1:

Newly published research indicates that the melt rate of the Antarctic Ice Sheet has tripled in the past ten years, threatening to accelerate sea level rise (The IMBIE team, 2018).

We had assumed that much of the deep ocean floor was still free of anthropogenic impacts, but research shows this is likely not the case, with toxic organic pollutants being detected in sediment cores from the Mariana Trench, the deepest ocean location in the entire world (Dasgupta *et al.*, 2018).

Chapter 4:

A reanalysis of climate models indicates that even greater projected global warming is likely by 2100 than the IPCC projections predict, suggesting that 'achieving any given global temperature stabilization target will require steeper greenhouse gas emissions reductions than previously calculated' (Brown and Caldeira, 2017, Abstract).

From their climate modeling, Seneviratne *et al.* (2018) point out that even keeping with the 1.5°C warming limit may have serious climatic impacts:

> 'Pursuing policies that are considered to be consistent with the 1.5°C aim will not completely remove the risk of global temperatures being much higher or of some regional extremes reaching dangerous levels for ecosystems and societies over the coming decades' (Abstract).

Since the issue of how much energy humanity really needs is so critical to the OWSP, we discuss the following paper at some length. The title of Grubler *et al.*'s (2018) paper is: 'A low energy demand scenario for meeting the 1.5°C target and sustainable development goals without negative emission technologies.' Their abstract states, 'Our scenario meets the 1.5°C climate target as well as many sustainable development goals, without relying on negative emission technologies.' Starting with our critique, this scenario does meet 'many sustainable development goals' but falls far short of creating equity between the global South and North, leaving the global South in a state of relative energy poverty, while failing to generate the energy capacity to meet the challenges of climatic adaptation, especially for the global South, and while failing to reach a sufficient precautionary level of mitigation in the form of aggressive carbon sequestration from the atmosphere into the soil but especially the crust. Even with their implementation of projected state-of-the-science energy efficiencies in all sectors, the likely necessary level of primary energy consumption by 2050 will be greater than the present level of 18 TW (see discussion in chapter 4). We note that the carbon dioxide level in the atmosphere post-2050 is not given in this paper. We conclude that a more robust global transition to wind/solar energy with global North/South equity would better serve the interests of future generations.

Here is supporting evidence for our critique, drawing from Grubler *et al.*'s (2018) extensive Supplement (we converted EJ, GJ at 2050 to corresponding power units TW, kW respectively). By 2050, the final energy demand for the global North is 1.55 kW per capita (1.6 billion population assumed), while for the global South this demand is 0.63 kW per capita (7.6 billion population assumed) (Tables 3 and 24). The total primary energy consumption corresponds to 8.27 TW, while the final energy demand corresponds to 6.91 TW (Figure 15b). In their 2050 scenario, if the global South receives the same final energy demand per capita as the global North, the total demand would correspond to 14.3 TW (while primary consumption would correspond to 17 TW). Note that their total contingency reserve corresponds to 0.5 TW (Table 24). In their discussion preceding Table 31, Grubler *et al.* (2018) say, 'Sequestration into newly grown forested land and bioenergy plantations biomass also increases the carbon sink which leads to a net sink of CO_2 in the land sector of –4.3 $GtCO_2$/year by the end century'. In comparison, Hansen *et al.* (2011) estimate that 100 Pg of carbon could be sequestered from the atmosphere by reforestation from 2031–2080 leaving 350 ppm CO_2 in the atmosphere by 2100, equivalent to an average **sequestration** rate of 7.3 $GtCO_2$/year. Finally, the 2050 scenario predicted by Grubler *et al.* (2018) still results in roughly 3 million premature deaths per year from ambient air pollution compared to a cited natural level of 0.5 million per year.

The issue of stranded fossil fuel assets is revisited by Mercure *et al.* (2018), while Burke *et al.* (2018) argue that economic damages and global inequality driven by climate change will be significantly reduced by meeting UN mitigation targets. Alperovitz *et al.* (2017) present a creative plan to 'save the climate from burning the same way it saved the economy from depression: Buy out the companies behind the crisis' (their article lead), in other words, use quantitative easing, print money, and nationalize the energy industry to cancel out the huge financial burden of the stranded assets of fossil fuels needed to remain unburned, as they put it:

> 'Following the 2008 crisis, the equivalent of roughly $12.3 trillion was pumped into the global financial system to repair the balance sheets of commercial banks. The US government, through the Federal Reserve, created around $3.5 trillion of new money between late 2008 and 2014, an average of nearly $600 billion a year. Not a penny of this new money had to be "paid for" through taxes or borrowing. And despite the massive scale, the runaway inflation that every economics student is told "must" result from such money creation has not materialized' (Alperovitz *et al.*, 2017).

We take note of new developments in technologies of carbon sequestration from the atmosphere from von Hippel (2018) and Keith *et al.* (2018) with commentary by Tollefson (2018) who emphasizes the dramatic cost reduction. Perasso (2018) gives an update of the Iceland project to mineralize carbon dioxide pumped down into the basalt.

We have not previously discussed CO_2 emissions from global cement production, a small but not negligible source. Xi *et al.* (2016) state in their abstract: 'Calcination of carbonate rocks during the manufacture of cement produced 5% of global CO_2 emissions from all industrial process and fossil-fuel combustion in 2013', but that the reversal of this process, the sequestration of atmospheric CO_2 in carbonating cement materials from 1930 to 2013, offset 43% of the CO_2 emissions from production of cement over the same period, not including emissions associated with fossil fuel use during cement production. In addition, research efforts continue to not only reduce CO_2 emissions from global cement production (e.g., Imbabi *et al.*, 2012), but to transform this technology to a carbon sink from the atmosphere, for example by using steel slag waste as a CO_2 reactive component of concrete production (Serebrin, 2018).

'$1.6 Trillion Natural Gas Expansion Will Eliminate Any Chance Of Meeting Paris Carbon Goals' is the title of a post which highlights the pending plan by G20 countries (Hanley, 2018), citing the report by Oil Change International (2018). We remind readers that because of methane leakage to the atmosphere, natural gas has the highest GHG footprint.

We conclude this supplement to chapter 4 by taking note of an important initiative to shed a global spotlight on the Energy Charter Treaty which is apparently aimed at granting 'corporations in the energy sector enormous power to sue states at international investment tribunals for billions of dollars, for example, if a government decides to stop new oil or gas pipelines or to phase out coal' (ECT's Dirty Secrets, 2018).

Chapter 5:

A potentially important addition to agroecological science is 'Evolutionary Plant Breeding', i.e., the development of new varieties from old ones by 'creating plant populations by mixing seeds previously obtained by crossing different varieties and letting them evolve using them as a crop or to select the best plants' (Ceccarelli, 2018).

Following up our discussion of Roundup and the Monsanto Corporation, the Corporate Europe Observatory (2018) claims that, 'internal documents released in a lawsuit by cancer victims show how the chemical giant actively subverts science to promote its products and profits', and further that, 'Revelations include confirmation that the company hardly tested the real-world toxicity of its products, actively avoided pursuing studies which might show unwelcome results, and ghostwrote the studies of supposedly independent scientists. The documents also show Monsanto systematically attacked scientists whose research threatened their profits, as aptly summarized in a 2001 email by a Monsanto executive....'.

Chapter 8:

For an insightful update on the current conditions in Venezuela and the challenges her government faces see Shupak (2018), whose subtitle reads, 'The intent of the sanctions is clear: to inflict maximum pain on Venezuela so as to encourage the people of the country to overthrow the democratically elected government'.

Chapter 9:

As a supplement to our discussion of prefiguration, here is a sample of the eloquent discussion of this concept from Ecosocialist Horizons (2018):

> *'An important political principle now emerges — one that applies to the production of use-values for the sustenance of life, and also to the production of ways beyond capital. The potential for the given to contain the lineaments of what is to be may be called prefiguration... To prefigure is also to return to what is truly immortal in humanity — our limitless imagination'.*

References cited (not included in the Bibliography)

Alperovitz, G., Guinan, J. and Hanna, T. M. (2017). *'The Policy Weapon Climate Activists Need'*, The Nation, 26 April 2017, [Online]. Available at: https://www.thenation.com/article/the-policy-weapon-climate-activists-need/. [Accessed 16 June 2018].

Brown, P. T. and Caldeira, K. (2017). Greater future global warming inferred from Earth's recent energy budget, *Nature*, 552, pp. 45–50, doi: 10.1038/nature24672.

Burke, M., Davis, W. M. and Diffenbaugh, N.S. (2018). Large potential reduction in economic damages under UN mitigation targets, *Nature*, 557, pp. 549–553, doi.: 10.1038/s41586-018-0071-9.

Ceccarelli S. (2018). *'Stuffed or Starved? Evolutionary Plant Breeding Might Have the Answer',* Independent Science News, [Online]. Available at: https://www.independentsciencenews.org/health/stuffed-or-starved-evolutionary-plant-breeding-might-have-the-answer/. [Accessed 13 June 2018].

Corporate Europe Observatory. (2018). March 1, What the Monsanto Papers tell us about corporate science, [Online]. Available at: https://corporateeurope.org/food-and-agriculture/2018/03/what-monsanto-papers-tell-us-about-corporate-science. [Accessed 13 June 2018].

Dasgupta, S., Peng, X., Chen, S. *et al.* (2018). Toxic anthropogenic pollutants reach the deepest ocean on Earth, *Geochem. Persp. Let.,* 7, pp. 22–26.

Ecosocialist Horizons. (2018). *'Ecosocialist Prefiguration',* [Online]. Available at: http://ecosocialisthorizons.com/prefiguration/. [Accessed 14 June 2018].

ECT's Dirty Secrets. (2018). *'What is the Energy Charter Treaty?'*, [Online]. Available at: https://www.energy-charter-dirty-secrets.org/. [Accessed 14 June 2018].

Grubler A., Wilson C., Bento, N. *et al.* (2018). A low energy demand scenario for meeting the 1.5°C target and sustainable development goals without negative emission technologies, *Nature Energy*, *Supplementary Information*, 3, 515–527, doi.: 10.1038/s41560-018-0172-6.

Hanley, S. (2018). *'\$1.6 Trillion Natural Gas Expansion Will Eliminate Any Chance Of Meeting Paris Carbon Goals',* CleanTechnica, 14 June 2018, [Online]. Available at: https://cleantechnica.com/2018/06/14/1-6-trillion-natural-gas-expansion-will-eliminate-any-chance-of-meeting-paris-carbon-goals/. [Accessed 18 June 2018].

Imbabi, M.S., Carrigan, C. and McKenna, S. (2012). Trends and developments in green cement and concrete technology, *International Journal of Sustainable Built Environment*, 1, 194–216.

The IMBIE team. (2018). Mass balance of the Antarctic Ice Sheet from 1992 to 2017, *Nature*, 558, pp. 219–222.

Keith, D.W., Holmes, G., St. Angelo, D. *et al.* (2018). A Process for Capturing CO_2 from the Atmosphere, *Joule*, 2, 1–22, doi.: 10.1016/j.joule.2018.05.006.

Kishore, N., Marqués, D., Mahmud, A. *et al.* (2018). Mortality in Puerto Rico after Hurricane Maria, *New England Journal of Medicine*, doi: 10.1056/NEJMsa1803972.

Mercure, J.-F., Pollitt, H., Viñuales, J.E. *et al.* (2018). Macroeconomic impact of stranded fossil fuel Assets, *Nature Climate Change*, doi: 10.1038/s41558-018-0182-1.

Oil Change International. (2018). *'Debunked: The G20 Clean Gas Myth'*, June 2018, [Online]. Available at: http://priceofoil.org/2018/06/11/report-g20-countries-set-to-invest-over-1-6-trillion-in-fossil-gas-jeopardizing-paris-climate-goals/. [Accessed 18 June 2018].

Perasso, V. (2018). *'Turning carbon dioxide into rock — forever'*. BBC World Service, 18 May 2018, [Online]. Available at: http://www.bbc.co.uk/news/world-43789527. [Accessed 13 June 2018].

Seneviratne, S.I., Rogelj, J., Séférian, R. *et al.* (2018). The many possible climates from the Paris Agreement's aim of 1.5°C warming, *Nature*, 558, pp. 41–49.

Serebrin, J. (2018). *'Montreal company pioneers carbon-negative concrete'*, Montreal Gazette, 29 April 2018, [Online]. Available at: http://montrealgazette.com/business/local-business/montreal-company-pioneers-carbon-negative-concrete. [Accessed 13 June 2018].

Shupak, G. (2018). *'Exonerating the Empire in Venezuela'*, Venezuelanalysis.com, 23 March 2018. [Online]. Available at: https://venezuelanalysis.com/ANALYSIS/13736. [Accessed 20 June 2018].

Tollefson, J. (2018). Price of sucking CO_2 from air plunges, N*ature*, 558, p. 173.

von Hippel, T. (2018). Thermal removal of carbon dioxide from the atmosphere: Energy requirements and scaling issues, *Climatic Change*, doi.: 10.1007/s10584-018-2208-0.

Xi, F., Davis, S.J., Crawford-Brown, D. *et al.* (2016). Substantial global carbon uptake by cement carbonation, *Nature Geoscience*, 9, pp. 880–883.

Bibliography

Abraham, J. (2015a). *'Changes in water vapor and clouds are amplifying global warming'*, The Guardian, 23 April 2015. [Online]. Available at: https://www.theguardian.com/environment/climate-consensus-97-per-cent/2015/apr/23/changes-in-water-vapor-and-clouds-are-amplifying-global-warming. [Accessed 24 August 2017].

Abraham, J. (2015b). *'Methane release from melting permafrost could trigger dangerous global warming'*, The Guardian, 13 October 2015. [Online]. Available at: https://www.theguardian.com/environment/climate-consensus-97-per-cent/2015/oct/13/methane-release-from-melting-permafrost-could-trigger-dangerous-global-warming. [Accessed 18 August 2017].

Abramsky, K. (ed.) (2010). *Sparking a worldwide energy revolution,* AK Press, Oakland.

Afanasyev, V. (1978). *Marxist Philosophy: A Popular Outline*, Foreign Language Publ. House, Moscow.

Ahmed, N. (2014). *'UN: Only Small Farmers and Agroecology Can Feed the World'*, 26 September 2014. [Online]. Available at: https://permaculturenews.org/2014/09/26/un-small-farmers-agroecology-can-feed-world/. [Accessed 23 August 2017].

Ahmed, N. (2015). *'The Anglo-American empire is preparing for resource war'*, Middle East Eye, 7 July 2015. [Online]. Available at: http://www.middleeasteye.net/columns/anglo-american-empire-preparing-resource-war-1170119289. [Accessed 23 August 2017].

Ahmed, N. (2016). *'This Could Be the Death of the Fossil Fuel Industry—Will the Rest of the Economy Go With It?'*, Truthout, 30 April 2016. [Online]. Available at: http://www.truth-out.org/news/item/35817-we-could-be-witnessing-the-death-of-the-fossil-fuel-industry-will-it-take-the-rest-of-the-economy-down-with-it. [Accessed 25 August 2017].

Ahmed, N. (2017a). *'Brace for Oil, Food and Financial Crash of 2018'*, Observer, 20 January 2017. [Online]. Available at: http://observer.com/2017/01/brace-for-the-oil-food-and-financial-crash-of-2018/. [Accessed 18 August 2017].

Ahmed, N.M. (2017b). *Failing States, Collapsing Systems: BioPhysical Triggers of Political Violence*, Springer, Berlin.

Akbari, H., Menon, S. and Rosenfeld, A. (2009). Global Cooling: Increasing World-wide Urban Albedos to Offset CO_2, *Climatic Change,* 94, pp. 275–286.

Alexander, M. (2010). *The new Jim Crow: Mass incarceration in the age of colorblindness*, New Press, New York.

Alexander, N. (ed.) (2014). *Infrastructure: for people or for profit?,* Heinrich Boll Foundation and Latindadd.

Alexandratos, N. and Bruinsma J. (2012). *'World agriculture towards 2030/2050: the 2012 revision'*, ESA Working Paper No. 12-03, [Online]. Available at: http://www.fao.org/docrep/016/ap106e/ap106e.pdf. [Accessed 27 January 2018].

Allen, T.W. (1997). *The Invention of the White Race (Two volumes, 2nd Edition, 2012, with introduction by Jeffrey B. Perry),* Verso Books, Brooklyn, N.Y.

Allred, B.W., Smith, K. W., Twidwell, D., *et al.* (2015). Ecosystem services lost to oil and gas in North America, *Science*, 348 (6233), pp. 401–402.

Alperovitz, G. (2013). *What Then Must We Do?,* Chelsea Green Publishing, White River Junction, Vermont.

Alperovitz, G. (2016). *'Socialism in America is Closer Than You Think'*, The Nation, 11 February 2016. [Online]. Available at: https://www.thenation.com/article/socialism-in-america-is-closer-than-you-think/. [Accessed 26 January 2018].

Altieri, M.A. (1995). *Agroecology: The Science of Sustainable Agriculture,* Westview Press, Boulder.

Altieri, M. (2016). *'Cuba's Sustainable Agriculture at Risk After U.S.-Cuba Relations Thaw'*. The New Republic, 26 March 2016. [Online]. Available at: https://newrepublic.com/article/132055/cubas-sustainable-agriculture-risk-us-cuba-relations-thaw. [Accessed 23 August 2017].

Altvater, E. (1994). 'Ecological and economic modalities of time and space', in O'Connor, M. (ed.), *Is Capitalism Sustainable?*, Guilford Press, New York, pp. 76–90.

Amelinckx, A. (2015). *'Even Without a Drought, We're Depleting Groundwater at an Alarming Pace'*, Modern Farmer, 30 July 2015. [Online]. Available at: http://modernfarmer.com/2015/07/ogallala-aquifer-depletion/. [Accessed 20 August 2017].

American Lung Association. (2014). *'Ozone Pollution'*. [Online]. Available at: http://www.stateoftheair.org/2014/key-findings/ozone-pollution.html?referrer=https://www.google.com/. [Accessed 18 August 2017].

Anderson, A. R. (Director). (2013). *Tierralismo* [DVD]. Cuba: Icarus Films.

Anderson, K. (2015). 'Duality in climate science', *Nature Geoscience,* 8, pp. 898–900.

Angus, I. (2016). *Facing the Anthropocene: Fossil Capitalism and the Crisis of the Earth System*, Monthly Review Press, New York.

Angus, I. and Butler, S. (2011). *Too Many People?*, Haymarket Books, Chicago.

Anthony, K. W. (2009). *'Arctic Climate Threat—Methane from Thawing Permafrost',* Scientific American, December. [Online]. Available at: https://www.scientificamerican.com/article/methane-a-menace-surfaces/. [Accessed 26 January 2018].

Appunn, K., Bieler, F. and Wettengel, J. (2017). *'Germany's energy consumption and power mix in charts'*. 1 August. [Online]. Available at: https://www.cleanenergywire.org/factsheets/germanys-energy-consumption-and-power-mix-charts. [Accessed 23 August 2017].

Archer, D. (2012). *Global Warming: Understanding the Forecast, 2nd Edition*, John Wiley: Hoboken, N. J.

Archer, C. L. and Jacobson, M. Z. (2007). 'Supplying baseload power and reducing transmission requirements by interconnecting wind farms', *Journal of Applied Meteorology and Climatology*, 46, pp. 1701–1717.

Arcieri, M. (2016). 'Spread and Potential Risks of Genetically Modified Organisms', *Agriculture and Agricultural Science Procedia*, 8, pp. 552–559.

Aronoff, K. (2016). *'How to Socialize America's Energy'*. Dissent Magazine, Spring 2016. [Online]. Available at: https://www.dissentmagazine.org/article/energy-democracy-usa-socialize-renewable-public-private-cooperatives. [Accessed 23 August 2017].

Asadi, M., Kim, K.,Liu, C., *et al.* (2016). 'Nanostructured transition metal dichalcogenide electrocatalysts for CO_2 reduction in ionic liquid', *Science,* 353, pp. 467–470.

Asici, A. A. and Bünül, Z. (2012). 'Green New Deal: A Green Way out of the Crisis?, Environmental Policy and Governance', *Env. Pol. Gov.*, 22, pp. 295–306.

Aston University. (2015). *'Seawater greenhouses to bring life to the desert'*, Science Daily, 14 July 2015. [Online]. Available at: http://www.sciencedaily.com/releases/2015/07/150714083029.htm. [Accessed 23 August 2017].

Atkins, P. W. (1984). *The Second Law*, Scientific American Books, New York.

AtKisson, A. (2009). *'A global green new deal for climate, energy, and development'*, United Nations, Department of Economic and Social Affairs, December 2009. [Online]. Available at: https://sustainabledevelopment.un.org/content/documents/cc_global_green_new_deal.pdf. [Accessed 26 January 2018].

Aune, J. B. (2012). 'Conventional, Organic and Conservation Agriculture: Production and Environmental Impact', *Agroecology and Strategies for Climate Change, Sustainable Agriculture Reviews*, 8, pp. 149–166.

Ayres, R. U. (1997). 'Comments on Georgescu-Roegen', *Ecological Economics,* 22 (3), pp. 285–287.

Ayers, R. U. (1998). 'Eco-Thermodynamics: Economics and the Second Law', *Ecological Economics*, 26 (2), pp. 189–209.

Ayers, R. U. (1999). 'The Second Law, the Fourth Law, Recycling and Limits to Growth', *Ecological Economics,* 29 (3), pp. 473–483.

Azzam, A. (2012). *'Energy Consumption in U.S. Food System'*, Cornhusker Economics, University of Nebraska-Lincoln Extension, 31 October 2012. [Online]. Available at: http://agecon.unl.edu/cornhusker-economics/2012/Energy-Consumption-FoodSystem.pdf. [Accessed 23 August 2017].

Badgley, C., Moghtader, J., Quintero, E., *et al.* (2007). 'Organic Agriculture and the Global Food Supply. *Renewable Agriculture and Food Systems',* 22 (2), pp. 86–108.

Bagg, J., Johnson, A., and Cumming, J. (2017). *'Crosby, Texas, Chemical Plant Explodes Twice, Arkema Group Says'*, 31 August. [Online]. Available at: https://www.nbcnews.com/storyline/hurricane-harvey/harvey-danger-major-chemical-plant-near-houston-likely-explode-facility-n797581. [Accessed 28 January 2018].

Baraniuk, C. (2017). *'How vertical farming reinvents agriculture'*, BBC. 6 April. [Online]. Available at: http://www.bbc.com/future/story/20170405-how-vertical-farming-reinvents-agriculture. [Accessed 23 August 2017].

Barca, S. (2015). 'Greening the job: trade unions, climate change and the political ecology of labour', in Bryant, R.L. (ed.), *International Handbook of Political Ecology,* Edward Elgar Publishing Limited, Cheltenham, pp. 387–400.

Barclay, E. (2014). *'U.S. Lets 141 Trillion Calories Of Food Go To Waste Each Year'*, National Public Radio, 27 February. [Online]. Available at: http://www.npr.org/sections/thesalt/2014/02/27/283071610/u-s-lets-141-trillion-calories-of-food-go-to-waste-each-year. [Accessed 23 August 2017].

Barron, J. (1983). *'High Cost of Military Parts'*, The New York Times, 1 September. [Online]. Available at: http://www.nytimes.com/1983/09/01/business/high-cost-of-military-parts.html?pagewanted=all. [Accessed 28 August 2017].

Barrow, J. D. (1994). *The Origin of the Universe,* Basic Books, New York.

Barrow, J. D. and Tipler, F. J. (1988). *The Anthropic Cosmological Principle*, Oxford University Press, New York.

Battiston, S., Mandel, A., Monasterolo, I. *et al.* (2017). 'A climate stress-test of the financial system', *Nature Climate Change*, 7, pp. 283–288.

Battistoni, A. (2016). 'Socialism or whole foods: Luxemburgian answers to our climate crisis', in Scharenberg, E.S. (ed.), *Rosa Remix*, Rosa Luxemburg Stiftung, New York, pp. 70–74.

BBC Online (2015). *'Global conflicts 'cost 13% of world GDP'*, BBC, 17 June 2015. [Online]. Available at: http://www.bbc.com/news/world-33161837. [Accessed 23 August 2017].

Bearak, M. and Gamio, L. (2016). *'The U.S. foreign aid budget, visualized'*, The Washington Post, 18 October. [Online]. Available at: https://www.washingtonpost.com/graphics/world/which-countries-get-the-most-foreign-aid/. [Accessed 23 August 2017].

Behera, K. K., Alam, A., Vats, S., *et al.* (2012). 'Organic Farming History and Techniques', *Agroecology and Strategies for Climate Change, Sustainable Agriculture Reviews,* 8, pp. 287–328.

Benyus, J. (1997). *Biomimicry: Innovation Inspired by Nature,* Harper Collins, New York.

Bergström, L. and Kirchmann, H. (2016). 'Are the claimed benefits of organic agriculture justified?', *Nature Plants*, 2 (7), 16099, DOI: 10.1038/nplants.2016.99.

Bernal, J. D. (1971). Originally published in 1954. *Science in History, Vol. 2, The Scientific and Industrial Revolutions,* MIT Press, Cambridge.

Bernstein, J. (2016). *'Wage outcomes and macroeconomic conditions: what's the connection?'*, Center on Budget and Policy Priorities, 16 August. [Online]. Available at: http://jaredbernsteinblog.com/wp-content/uploads/2016/08/wg_macro_8_25.pdf. [Accessed 23 August 2017].

Bertalanffy, L. V. (1968). *General System Theory*, George Braziller, New York.

Bertsch, J. (n.d.). *'Morelly, (Étienne-Gabriel?)'*. [Online]. Available at: http://enlightenment-revolution.org/index.php/Morelly%2C_%28%C3%89tienne-Gabriel%3F%29. [Accessed 27 August 2017].

Bevan, M. W., Uauy, C., Brande, B. H., *et al.* (2017). 'Genomic innovation for crop improvement', *Nature,* 543, pp. 346–354.

Bhattacharya, T. (ed.) (2017). *Social Reproduction Theory: Remapping Class, Recentering Oppression*, Pluto Press, London.

Bianciardi, C., Tiezzi, E., and Ulgiati, S. (1993). 'Complete recycling of matter in the frameworks of physics, biology and ecological economics', *Ecological Economics*, 8, pp. 1–5.

Biel, R. (2016). *Sustainable Food Systems: The Role of the City,* UCL Press, London.

Biello, D. (2010). '*Genetically Modified Crop on the Loose and Evolving in the U.S. Midwest'*, Scientific American, 6 August. [Online]. Available at: https://www.scientificamericancom/article/genetically-modified-crop/. [Accessed 24 August 2017].

Billings, L. (2017). *'Q&A: Kim Stanley Robinson Explains How He Flooded Manhattan'*, 13 March. [Online]. Available at: https://www.scientificamerican.com/article/q-a-kim-stanley-robinson-explains-how-he-flooded-manhattan/. [Accessed 23 August 2017].

Biodiversity for a Livable Climate. (2015). *'Urban Eco-Restoration Series'*, 22 September. [Online]. Available at: http://myemail.constantcontact.com/Biodiversity-for-a-Livable-Climate-September-2015-Newsletter.html?soid=1119463932201&aid=wx26Ih0ODGw. [Accessed 20 August 2017].

Biomimicry Institute. (2015). '*Jube—2015 Biomimicry Global Design Challenge Finalist*.' [Video file]. [Online]. Available at: https://vimeo.com/140968606. [Accessed 29 January 2018].

Birchall, I. (2016). *The Spectre of Babeu*f (*Second Edition)*, Haymarket Books, Chicago.

Biswas, S. (2014). *'India's Dark History of Sterilization',* BBC News, 14 November. [Online]. Available at: http://www.bbc.com/news/world-asia-india-30040790. [Accessed 23 August 2017].

Black, E. (2008). *Internal Combustion: How Corporations and Governments Addicted the World to Oil and Derailed the Alternatives,* Dialog Press.

Blakers, A. (2017). *'Solar is now the most popular form of new electricity generation worldwide'*, 2 August. [Online]. Available at: https://theconversation.com/solar-is-now-the-most-popular-form-of-new-electricity-generation-worldwide-81678 [Accessed 26 January 2018].

Blasing, T. J. (2016). *'Recent Greenhouse Gas Concentrations'*, CDIAC (Carbon Dioxide Information Analysis Center). [Online]. (Updated April 2016). Available at: http://cdiac.ornl.gov/pns/current_ghg.html. [Accessed 18 August 2017].

Bloch, E. (1986). *The Principle of Hope (three volumes)*, MIT Press, Cambridge.

Bluegreen Alliance. (2017). *'Bluegreen Alliance website'*, [Online]. Available at: http://www.bluegreenalliance.org. [Accessed 23 August 2017].

BNEF. (2017). *'New Energy Outlook 2017'*, Bloomberg New Energy Finance. [Online]. Available at: https://about.bnef.com/new-energy-outlook/. [Accessed 23 August 2017].

Bock, M., Schmitt, J., Beck, J., *et al.* (2017). Glacial/interglacial wetland, biomass burning, and geologic methane emissions constrained by dual stable isotopic CH_4 ice core records', *Proceedings of the National Academy of Sciences (USA)*, 114 (29), pp. E5778-E5786.

Boggs, G. L. (2012). *The Next American Revolution: Sustainable Activism for the Twenty-first Century,* University of California Press, Berkeley.

Bonaiuti, M. (2012). 'Degrowth: Tools for a complex analysis of the multidimensional crisis', *Capitalism Nature Socialism,* 23 (1), pp. 30–50.

Bond, P. (2010). *'The bank loan that could break South Africa's back'*, 13 April. [Online]. Available at: http://mrzine.monthlyreview.org/2010/bond130410.html. [Accessed 18 August 2017].

Bond, P. (2015). *'Paris climate agreement: a terror attack on Africa'*, 17 December. [Online]. Available at: http://climateandcapitalism.com/2015/12/17/paris-climate-agreement-a-terror-attack-on-africa/. [Accessed 18 August 2017].

Bond, P. and Garcia, A. (eds.) (2015). *BRICS: An Anti-Capitalist Critique*, Haymarket, Chicago.

Boucher, D., Barclay, B., Lichtenberg, E., *et al.* (1993). 'Review of 'For the Common Good', *Capitalism Nature Socialism',* 4 (3), pp. 129–135.

Boudreaux, R. (1991). *'Cuba Curbs Reliance on Soviets: Trade: Benefactor's upheaval has a worried Castro seeking economic self-sufficiency'*, Los Angeles Times, 18 August. [Online]. Available at: http://articles.latimes.com/1991-08-18/news/mn-1518_1_soviet-union. [Accessed 28 August 2017].

Bountiful Energy Blog (2017). [Online]. Available at: http://bountifulenergy.blogspot.com. [Accessed 27 August 2017].

Bowring, B. (2008). *The Degradation of the International Legal Order?*, Routledge, New York.

BP. (2017). *'Statistical Review of World Energy'*, BP. [Online]. Available at: http://www.bp.com/en/global/corporate/energy-economics/statistical-review-of-world-energy.html. [Accessed 22 August 2017].

Bradsher, K. (2017). *'China Looks to Capitalize on Clean Energy as U.S. Retreats'*, New York Times, 5 June. [Online]. Available at: https://www.nytimes.com/2017/06/05/business/energy-environment/china-clean-energy-coal-pollution.html?mcubz=3. [Accessed 23 August 2017].

Breitburg, D., Levin, L.A., Oschlies, A., *et al.* (2018). Declining oxygen in the global ocean and coastal waters, *Science*, 359 (6371), eaam7240, DOI: 10.1126/science.aam7240.

Brescia, S. (ed.) (2017). *Fertile Ground: Scaling Agroecology from the Ground Up*, Food First Books, Oakland.

Brink, S. (2017). *'What Country Spends the Most (And Least) On Health Care Per Person'*, NPR, 20 April. [Online]. Available at: http://www.npr.org/sections/goatsandsoda/2017/04/20/524774195/what-country-spends-the-most-and-least-on-health-care-per-person. [Accessed 23 August 2017].

Bruinsma, J. (ed.). (2003). *'World agriculture, towards 2030/2050'*, United Nations Organisation for Food and Agriculture (FAO), Rome. [Online]. Available at: http://www.fao.org/3/a-y4252e.pdf. [Accessed 24 April 2017].

Buckley, C. (2017). *'Xi Jinping is Set for a Big Gamble with China's Carbon Trading Market'*, New York Times, 23 June. [Online]. Available at: https://www.nytimes.com/2017/06/23/world/asia/china-cap-trade-carbon-greenhouse.html. [Accessed 26 January 2018].

Buddhiraju, S., Santhanam, P., Fan, S. (2018). 'Thermodynamic limits of energy harvesting from outgoing thermal radiation', *Proc Natl Acad Sci (USA)*, DOI: 10.1073/pnas.1717595115.

Burkett, P. (1998). 'A Critique of Neo-Malthusian Marxism: Society, Nature, and Population', *Historical Materialism*, 2 (1), pp. 118–42.

Burkett, P. (2003). 'The Value Problem in Ecological Economics, Lessons from the Physiocrats and Marx', *Organization & Environment*, 16 (2), pp. 137–167.

Burkett, P. (2005). 'Entropy in Ecological Economics: A Marxist Intervention', *Historical Materialism*, 13 (1), pp. 117–152.

Burkett, P. (2006). *Marxism and Ecological Economics*, Brill, Leiden.

Buxton, N. (2015). *'The elephant in Paris—the military and greenhouse gas emissions'*, New Internationalist, 19 November. [Online]. Available at: https://newint.org/blog/2015/11/19/the-military-and-greenhouse-gas-emissions. [Accessed 23 August 2017].

Caffentzis, G. (2013). 'From the Grundrisse to Capital and Beyond: Then and Now', in, Bellofiore, R., Starosta, G., and Thomas, P.D. (eds.), *Marx's Laboratory*, Haymarket, Chicago, pp. 265–281.

Camazine, S., Deneubourg, J. L., Franks, N. R., *et al.* (2001). *Self-organization in biological systems*, Princeton University Press, Princeton.

Cameron, D., Osborne, C., Horton, F. R. S., *et al.* (2015). 'A sustainable model for intensive agriculture', December. [Online]. Available at: http://grantham.sheffield.ac.uk/wp-content/uploads/2015/12/A4-sustainable-model-intensive-agriculture-spread.pdf. [Accessed 23 August 2017].

Campaign Against Climate Change. (2017). *'One Million Climate Jobs'*. [Online]. Available at: https://www.cacctu.org.uk/climatejobs. [Accessed 23 August 2017].

Canada Crude Oil Production. (2017). [Online]. Available at: https://ycharts.com/indicators/canada_crude_oil_production. [Accessed 23 August 2017].

Canning, P., Charles, A., Huang, S., *et al.* (2010). *'Energy Use in the U.S. Food System'*, USDA, Economic Research Report Number 94, March. [Online]. Available at: http://web.mit.edu/dusp/dusp_extension_unsec/reports/polenske_ag_energy.pdf. [Accessed 10 August 2017].

Cao, J., Cohen, A., Hansen, J., *et al.* (2016). 'China-U.S. cooperation to advance nuclear power', *Science*, 353, pp. 547–548.

Cao, L. and Caldeira, K. (2010). 'Atmospheric carbon dioxide removal: long-term consequences and commitment', *Environ. Res. Lett.*, 5, DOI: 10.1088/1748-9326/5/2/024011.

Capuchin Soup Kitchen. (2017). [Online]. Available at: http://www.cskdetroit.org. [Accessed 22 August 2017].

Carbon Tracker. (2011). *'The Unburnable Carbon – Are the World's Financial Markets Carrying a Carbon Bubble?'*, Carbon Tracker Initiative, London,

13 July. [Online]. Available at: https://www.carbontracker.org/wp-content/uploads/2014/09/Unburnable-Carbon-Full-rev2-1.pdf. [Accessed 26 January 2018].

Carbon Tracker. (2015). *'The $2 trillion stranded assets danger zone: How fossil fuel firms risk destroying investor returns'*, Carbon Tracker Initiative, London. [Online]. Available at: https://www.carbontracker.org/wp-content/uploads/2015/11/CAR3817_Synthesis_Report_24.11.15_WEB2.pdf. [Accessed 26 January 2018].

Carbon Trade Watch. (2015). *'Paths Beyond Paris: Movements, Action and Solidarity Towards Climate Justice,* [Online]. Available at: http://www.carbontradewatch.org/downloads/publications/PathsBeyondParis-EN.pdf. [Accessed 23 August 2017].

Cardwell, D. S. L. (1971). *From Watt to Clausius*, Iowa State University Press, Ames, Iowa.

CARI. (2017). *'Data: Chinese Loans to Africa'*, China Africa Research Initiatives. [Online]. Available at: http://www.sais-cari.org/data-chinese-loans-and-aid-to-africa/. [Accessed 28 August 2017].

Carrington, D. (2016). *'Growing underground: the fresh herbs sprouting beneath Londoners' feet'*, The Guardian. 26 April. [Online]. Available at: https://www.theguardian.com/environment/2016/apr/26/growing-underground-the-fresh-herbs-sprouting-beneath-londoners-feet. [Accessed 23 August 2017].

Carson, R. (1962). *Silent Spring,* Houghton Mifflin, Boston.

Carter, R. (2017). *'Air pollution is killing wildlife and people'*, The Guardian, 14 June. [Online]. Available at: https://www.theguardian.com/environment/2017/jun/14/air-pollution-is-killing-wildlife-and-people. [Accessed 23 August 2017].

Cartlidge, E. (2017). *'Fusion energy pushed back beyond 2050'*, BBC News, 11 July. [Online]. Available at: http://www.bbc.com/news/science-environment-40558758. [Accessed 23 August 2017].

CASSE (2017). Center for the Advancement of the Steady State Economy. [Online]. Available at: http://www.steadystate.org/. [Accessed 23 August 2017].

CBD. (2017). Center for Biological Diversity. [Online]. Available at: http://www.biologicaldiversity.org. [Accessed 24 August 2017].

CDC. (2017). *'Global WASH Fast Facts'*, Global Water, Sanitation, & Hygiene (WASH). Centers for Disease Control and Prevention. [Online]. Available at:

https://www.cdc.gov/healthywater/global/wash_statistics.html. [Accessed 26 January 2018].

Ceballos, G., Ehrlich, P. R., Barnosky, A. D., *et al.* (2015). 'Accelerated modern human–induced species losses: Entering the sixth mass extinction', *Science Advances,* 1(5), e1400253, DOI: 10.1126/sciadv.1400253.

Ceballos, G., Ehrlich, P. R. and Dirzo, R. (2017). 'Biological annihilation via the ongoing sixth mass extinction signaled by vertebrate population losses and declines', *Proceedings of the National Academy of Sciences (U.S.A.),* 114 (30), pp. E6089-E6096.

Chadburn, S. E., Burke, E. J., Cox, P. M., *et al.* (2017). 'An observation-based constraint on permafrost loss as a function of global warming', *Nature Climate Change*, 7, pp. 340–344.

Chamberlin, T.W. (1940). *'Rainfall Maps of Cuba'*, Monthly Weather Review, pp. 4–10. [Online]. Available at: https://docs.lib.noaa.gov/rescue/mwr/068/mwr-068-01-0004.pdf. [Accessed 8 July 2017].

Chandler, A. (2016). *'Where the Poor Spend More Than 10 Percent of Their Income on Energy'*, The Atlantic, 8 June 2016. [Online]. Available at: https://www.theatlantic.com/business/archive/2016/06/energy-poverty-low-income-households/486197/. [Accessed 23 August 2017].

Chapin III, F. S., Matson, P. A. and Mooney, H. A. (2002). *Principles of Terrestrial Ecosystem Ecology,* Springer-Verlag, New York.

Charles, C., Gerasimchuk, I., Bridle, R., *et al.* (2013). *'Biofuels—At What Cost? A review of costs and benefits of EU biofuel policies'*, The International Institute for Sustainable Development, April. [Online]. Available at: http://www.iisd.org/gsi/sites/default/files/biofuels_subsidies_eu_review.pdf. [Accessed 8 August 2017].

Chen, B., Dong, L., Liu, X., *et al.* (2016). 'Exploring the possible effect of anthropogenic heat release due to global energy consumption upon global climate: a climate model study', *International Journal of Climatology*, 36 (15), pp. 4790–4796.

Cheng, V. K. M. and Hammond, G. P. (2017). 'Life-cycle energy densities and land-take requirements of various power generators: A UK perspective', *Journal of the Energy Institute,* 90, pp. 201–213.

Chirps. (2017). *'Chirps Chips'*. [Online]. Available at: http://chirpschips.com. [Accessed 28 August 2017].

Choi, C. Q. (2015). *'Air Pollution Kills More than 3 Million People Globally Every Year'*, Live Science, 16 September. [Online]. Available at: https://www.livescience.com/52189-air-pollution-kills-millions-people-yearly.html. [Accessed 23 August 2017].

CIA. (2017a). *'The World Factbook, Urbanization'*. [Online]. Available at: https://www.cia.gov/library/publications/the-world-factbook/fields/2212.html. [Accessed 23 August 2017].

CIA. (2017b). *'The World Factbook, Population Growth Rate'*. [Online]. Available at: https://www.cia.gov/library/publications/the-world-factbook/rankorder/2002rank.html. [Accessed 23 August 2017].

Clack, C. T. M., Qvist, S. A., Apt, J., *et al.* (2017). 'Evaluation of a proposal for reliable low-cost grid power with 100% wind, water, and solar', *Proc. Natl Acad. Sci. (USA)*, 114 (26), pp, 6722–6727.

Clay, J. (2004). *World Agriculture and the Environment: A Commodity-by-Commodity Guide to Impacts and Practices,* Island Press, Chicago.

Cleaver, H. (1992). 'The Inversion of Class Perspective in Marxian Theory: From Valorisation to Self–Valorisation', in Bonefeld, W., *et al.* (eds.), *Open Marxism, Vol. II*, Pluto Press, London, pp. 106–144.

Cleaver, H. (2011). 'Work Refusal and Self-Organisation', in Nelson, A. and Timmerman, F. (eds.), *Life Without Money: Building Fair and Sustainable Economies*, Pluto Press, London, pp. 47–69.

Cleaver, H. and De Angelis, M. (1993). *'An Interview with Harry Cleaver'*. [Online]. https://la.utexas.edu/users/hcleaver/InterviewwithHarryCleaver.html. [Accessed 28 August 2017].

Cleveland, C. J. and Ruth, M. (1997). When, Where, and By How Much Do Biophysical Limits Constrain the Economic Process? A survey of Nicholas Georgescu-Roegen's Contribution to Ecological Economics, *Ecological Economics*, 22, pp. 203–223.

Climate Action Tracker. (2015). *'INDCs lower projected warming to 2.7 °C: significant progress but still above 2° C'*, 1 October. [Online]. Available at: http://climateactiontracker.org/news/224/indcs-lower-projected-warming-to-2.7c-significant-progress-but-still-above-2c-.html. [Accessed 18 August 2017].

Climate Interactive. (2017). *'Climate Scoreboard. UN Climate Pledge Analysis'*. [Online]. Available at: https://www.climateinteractive.org/programs/scoreboard/. [Accessed 18 August 2017].

Climatemp. (2017). *'Rainfall/Precipitation for Havana, Cuba'*. [Online]. Available at: http://www.havana.climatemps.com/precipitation.php. [Accessed 28 August 2017].

Climate Refugees. (2017). [Online]. Available at: http://climaterefugees.com. [Accessed 23 August 2017].

CO2Earth. (2017). *'Earth's CO_2 Home Page'*. [Online]. Available at: https://www.co2.earth/. [Accessed 24 August 2017].

Coady, D., Parry, I., Sears, L., *et al.* (2015). *'How Large Are Global Energy Subsidies?, IMF Working Paper'*. Fiscal Affairs Department. [Online]. Available at: https://www.imf.org/external/pubs/ft/wp/2015/wp15105.pdf. [Accessed 26 January 2018].

Cobb, T. D. (2009). *Reclaiming Our Food: How the Grassroots Food Movement Is Changing the Way We Eat,* Storey Publishing, North Adams, Massachusetts.

Commane, R., Lindaas, J., Benmergui, J., *et al.* (2017). 'Carbon dioxide sources from Alaska driven by increasing early winter respiration', *Proc. Natl Acad. Sci. (USA)*, 114 (21), pp. 5361–5366.

Commoner, B. (1971). *The Closing Circle: Nature, Man, and Technology*, Alfred A. Knopf, New York.

Commoner, B. (1976). *The Poverty of Power*, Alfred A. Knopf, New York.

Commoner, B. (1990). *Making Peace with the Planet*, Pantheon, New York.

Conner, C. D. (2005). *A People's History of Science*, Nation Books, New York.

Convention on Biological Diversity. (2017). *'List of Parties'*. [Online]. Available at: https://www.cbd.int/information/parties.shtml. [Accessed 28 August 2017].

Cook, L. and Cherney, E. (2017). *'Get Ready for Peak Oil Demand'*, The Wall Street Journal, 26 May. [Online]. Available at: https://www.wsj.com/articles/get-ready-for-peak-oil-demand-1495419061. [Accessed 23 August 2017].

Cope, S. (2016). *'Why So Many Cubans 'Grow Their Own Food'*. City Lab, 29 January. [Online]. Available at: https://www.citylab.com/life/2016/01/cuba-urban-agriculture/433884/. [Accessed 23 August 2017].

Coveney, P. and Highfield, R. (1990). *The Arrow of Time,* Fawcett Columbine, New York.

Cox, S. (2016). *'Enough with the vertical farming fantasies: There are still too many unanswered questions about the trendy practice'*, Salon, 17 February. [Online]. Available at: http://www.salon.com/2016/02/17/enough_with_the_vertical_farming_partner/. [Accessed 23 August 2017].

Crawford, N.C. (2017). *'Costs of War'*, Watson Institute, Brown University, November, [Online]. Available at http://watson.brown.edu/costsofwar/figures/2017/us-budgetary-costs-post-911-wars-through-fy2018-56-trillion, [Accessed 24 January 2018].

Cressey, D. (2015). *'Widely Used Herbicide Linked to Cancer'*, Scientific American–Nature, 25 March. [Online]. Available at: https://www.scientificamerican.com/article/widely-used-herbicide-linked-to-cancer/. [Accessed 26 January 2018].

Crick. (2017). *'Crick Flours'*. [Online]. Available at: https://www.cricketflours.com/how-to-make-cricket-flour/. [Accessed 26 January 2018].

Crill, P.M. and Thornton, B.F. (2017). Whither methane in the IPCC process?, *Nature Climate Change*, 7, pp. 678–680.

Crowther, T. W., Todd-Brown, K. E. O., Rowe, C. W., *et al.* (2016). Quantifying global soil carbon losses in response to warming, *Nature,* 540, pp. 104–108.

Crutzen, P. J. and Stoermer, F. F. (2000). The "Anthropocene", *IGBP Newsletter*, 41, pp. 12–14.

Cullen, J. M. and Allwood, J. M. (2010). Theoretical efficiency limits for energy conversion devices, *Energy*, 35, pp. 2059–2069.

Cullen, J. M., Allwood, J. M., and Borgstein, E. H. (2011). 'Reducing energy demand: what are the practical limits?, *Environ. Sci. Technol.*', 45, pp. 1711–1718.

Daly, H. E. (1991). *Steady-State Economics* (2nd edition), Island Press, Washington, DC.

Daly, H. E. (1992). 'Is the entropy law relevant to the economics of natural resource scarcity? — yes, of course it is!', *Journal of Environmental Economics and Management,* 23 (1), pp. 91–95.

Daly, H. E. (1995). 'Reply to Mark Sagoff's Carrying capacity and ecological economics', *BioScience,* 45 (9), pp. 621–624.

Daly, H. E. and Cobb, J. B. (1989). *For the Common Good*, Beacon Press, Boston.

Daly, H. E. and Umana, A. F. (eds). (1981). *Energy, Economics, and the Environment,* AAAS Selected Symposium 64, Boulder, Colorado: Westview Press, Boulder.

Davies, J. H. and Davies, D. R. (2010). 'Earth's surface heat flux', *Solid Earth*, 1(1), pp. 5–24.

Davies, P. C. W. (1977). *The Physics of Time Asymmetry*, California: University of California Press, Berkeley.

Davis, D. (2002). *When Smoke Ran Like Water*, Basic Books, New York.

Davis, J. (2000). *'The End of Value'*, Marxism 2000 Conference, September. [Online]. Available at: http://www.gocatgo.com/texts/eov.html. [Accessed 28 August 2017].

Davis, J., Hirschl, T. and Stack, M. (eds). (1997). *Cutting Edge*, Verso, London.

Davis, M. (2006). *Planet of slums,* Verso, London.

Davis, M. (2010). 'Who will build the Ark?', *New Left Review,* 61, pp. 29–46.

Davis, S. J., Caldeira, K. and Matthews, H. D. (2010). 'Future CO_2 emissions and climate change from existing energy infrastructure', *Science,* 329, pp. 1330–1333.

Dean, J. (2012). *The Communist Horizon*, Verso, London.

DeHaan, L. R., Cox, T. S., Van Tassel, D. L., *et al.* (2007). 'Perennial Grains', in Scherr, S. J. and McNeely, J. A. (eds.), *Farming with Nature: The Science and Practice of Ecoagriculture*, Island Press, Washington D.C., pp. 61–82.

Delucchi, M. A. and Jacobson, M. Z. (2011). 'Providing all global energy with wind, water, and solar power, Part II: Reliability, system and transmission costs, and policies', *Energy Policy,* 39. pp. 1170–1190.

de Nijs, B. (2016). *'Are insects the future crop for the vertical farmer?'*, 27 June. [Online]. http://www.hortidaily.com/article/27099/Are-insects-the-future-crop-for-the-vertical-farmer. [Accessed 28 August 2017].

De Schutter, O. (2012). 'Agroecology, a Tool for the Realization of the Right to Food', *Agroecology and Strategies for Climate Change, Sustainable Agriculture Reviews,* 8, pp. 1–16.

Despommier, D. (2016). *The Vertical Farm: Feeding the World in the 21st Century,* St. Martin's Press, New York.

Detroit Agriculture. (2017). *'Keep Growing Detroit'*. [Online]. Available at: http://detroitagriculture.net. [Accessed 24 August 2017].

Diamond, J. (1999). *Guns, Germs & Steel,* W. W. Norton & Company, New York.

Diamond, J. (2005). *Collapse: How Societies Choose to Fail or Succeed,* Viking Penguin, New York.

Díaz-Briquets, S. and Pérez-López, J. F. (1995). *'The Special Period and the Environment'*, The Association for the Study of the Cuban Economy (ASCE), 30 November. [Online]. Available at: http://www.ascecuba.org/asce_proceedings/the-special-period-and-the-environment/. [Accessed 28 August 2017].

Dich, J., Zahm, S. H., Hanberg, A., *et al.* (1997). 'Pesticides and cancer', *Cancer Causes Control,* 8(3), pp. 420–443.

Dietz, S., Bowen, S., Dixon, C., *et al.* (2016). 'Climate value at risk of global financial assets', *Nature Climate Change*, 6, pp. 676–679.

Dinerstein, A. (2015). *The Politics of Autonomy in Latin America*, Palgrave Macmillan, N.Y.

Dinerstein, A. (2016a). *'Seminar in contemporary marxist theory: discussing utopia in 21st century capitalism',* 29 November. [Online]. Available at: https://www.kcl.ac.uk/sspp/departments/european-studies/eventrecords/2016-17/marxseminar3.aspx. [Accessed 28 August 2017].

Dinerstein, A. (2016b). *'Connecting the new and the old 'labour' struggles: Marxism, social reproduction and concrete utopia', Marxism and Contemporary South Asia: Relevance and Issues, India International Centre,* New Delhi, 11 November. [Online]. Available at: http://insurgentscripts.org/international-conference-on-marxism-and-contemporary-south-asia-issues-and-relevance/. [Accessed 26 January 2018].

Dobson, M. (2018). *'The Radical Paris Agreement'*, Jacobin, 5 January. [Online]. Available at: https://www.jacobinmag.com/2018/01/paris-climate-agreement-global-warming-trump. [Accessed 24 January 2018].

Dougherty-Choux, L. (2015). *'The Costs of Climate Adaptation, Explained in 4 Infographics*', World Resources Institute, 23 April. [Online]. Available at: http://www.wri.org/blog/2015/04/costs-climate-adaptation-explained-4-infographics. [Accessed 28 August 2017].

Dowdall, C. and Klotz, R. (2016). *Pesticides and Global Health: Understanding Agrochemical Dependence and Investing in Sustainable Solutions*, Routledge, New York.

Drewnowski, A., Rehm, C. D., Martin, A., *et al.* (2015). 'Energy and nutrient density of foods in relation to their carbon footprint', *Am J Clin Nutr*, 101, pp. 184–191.

Drijfhout, S., Bathiany, S., Beaulieu, C., *et al.* (2015). 'Catalogue of abrupt shifts in Intergovernmental Panel on Climate Change climate models', *Proc. Natl. Acad. Sci. (U.S.A.),* 112(43), pp. E5777–E5786.

Dryzek, J. S. (1994). 'Ecology and discursive democracy: beyond liberal capitalism and the administrative state', in O'Connor, M. (ed.)., *Is Capitalism Sustainable?*, Guilford Press, New York, pp. 176–197.

Dyer-Witheford, N. (1999). *Cyber-Marx, Cycles and Circuits of Struggle in High Technology Capitalism*, University of Illinois Press, Urbana and Chicago.

Dyer-Witheford, N. (2015). *Cyber-Proletariat,* Pluto Press, London.

Dyke, C. (1988). *The Evolutionary Dynamics of Complex Systems,* Oxford University Press, New York.

Ecosocialist Horizons. (2016). *'The Cry of Mother Earth: Call to the First Ecosocialist International',* 30 November. [Online]. Available at: http://ecosocialisthorizons.com/2016/11/the-cry-of-mother-earth-call-to-the-first-ecosocialist-international/. [Accessed 26 August 2017].

Ehrlich, P. R. (1968). *The Population Bomb*. Ballantine Books, New York.

Ellis, R. (2016). *'UNICEF: Air pollution kills 600,000 children yearly'*, CNN. [Online]. (Updated 31 October 2016). Available at: http://www.cnn.com/2016/10/30/health/air-pollution-children-unicef/index.html. [Accessed 28 August 2017].

Enerdata. (2011). *'World energy expenditures have more than doubled in 20 years'*, 28 November. [Online]. Available at: https://www.enerdata.net/publications/executive-briefing/world-energy-expenditures.html. [Accessed 27 August 2017].

Enerdata (2017). *'Total energy consumption (in 2016)'*. [Online]. Available at: https://yearbook.enerdata.net/total-energy/world-consumption-statistics.html. [Accessed 27 August 2017].

Energy Justice. (2017). *Energy Justice Network*. [Online]. Available at: http://www.energyjustice.net/. [Accessed 2 August 2017].

Energy Justice EU. (2017). *European Energy Justice Network*. [Online]. Available at: https://www.energyjustice.eu/. [Accessed 2 August 2017].

Engel-Di Mauro, S. (2014). *Ecology, Soils, and the Left: An Eco-Social Approach*, Palgrave Macmillan, New York.

Engels, F. (1987). *Karl Marx, Frederick Engels, Collected Work. Vol. 25*, International Publishers, New York.

Erbentraut, J. (2015). *'Why Vertical Farming Could Be On The Verge Of A Revolution — and What's Keeping It Down'*, The Huffington Post. [Online]. (Updated 29 March 2015). Available at: http://www.huffingtonpost.com/2015/03/11/vertical-farm-industry_n_6818402.html. [Accessed 28 August 2017].

Ergas, C. (2013). *'Cuban Urban Agriculture as a Strategy for Food Sovereignty'*, Monthly Review, 64 (10), March, [Online]. Available at: https://monthlyreview.org/2013/03/01/cuban-urban-agriculture-as-a-strategy-for-food-sovereignty. [Accessed 28 August 2017].

European Business Council for Sustainable Energy. (2017). *Trans-Mediterranean Renewable Energy Cooperation (TREC)*. [Online]. Available at: http://www.e5.org/cooperations-spin-offs/trec/. [Accessed 27 August 2017].

European Commission. (2008). *'Urbanization: 95% Of The World's Population Lives On 10% Of The Land'*, European Commission Joint Research Centre (JRC). 19 December. [Online]. Available at: www.sciencedaily.com/releases/2008/12/081217192745.htm. [Accessed 28 August 2017].

Fanelli, C. and Noonan, J. (2017). 'Capital and organized labour', in Schmidt, I. and Fanelli, C. (eds.), *Reading 'Capital' Today*, Pluto Press, London, pp. 138–159.

FAO. (2012). 'World agriculture, towards 2030/2050. The 2012 Revision', in Alexandratos, N. and Bruinsma, J. (eds.). *United Nations Organisation for Food and Agriculture,* Rome.

FAO. (2014). *'Pesticides Use'*, Food and Agriculture Organization of the United Nations. [Online]. Available at: http://www.fao.org/faostat/en/#data/RP. [Accessed 24 August 2017].

FAO. (2015a). *'World fertilizer trends and outlook to 2018'*, Food and Agriculture Organization of the United Nations. [Online]. Available at: http://www.fao.org/3/a-i4324e.pdf. [Accessed 29 August 2017].

FAO. (2015b). *'Agroecology for food security and nutrition'*, Proceedings of the FAO International Symposium on Agroecology for Food Security and Nutrition, 18–19 September 2014, Rome, Food and Agriculture Organization of the United Nations. [Online]. Available at: http://www.fao.org/3/a-i4729e.pdf. [Accessed 22 February 2018].

FAO. (2017a). *'SAVE FOOD: Global Initiative on Food Loss and Waste Reduction'*, Food and Agriculture Organization of the United Nations. [Online]. Available at: http://www.fao.org/save-food/resources/keyfindings/en/. [Accessed 24 August 2017].

FAO. (2017b). *'Global Action on Pollination Services for Sustainable Agriculture'*, Food and Agriculture Organization of the United Nations. [Online]. Available at: http://www.fao.org/pollination/en/. [Accessed 24 August 2017].

Farrell, J. (2016). *Beyond Sharing — How Communities Can Take Ownership of Renewable Power,* Institute for Local Self-Reliance, Washington, DC.

Farrell, J. (2017). *'Energy and Policy Institute Exposes Three Ways Electric Utilities Stomp Innovation and Competition'*, Institute for Local Self-Reliance, 15 August. [Online]. Available at: https://ilsr.org/energy-and-policy-institute-exposes-three-ways-electric-utilities-stomp-innovation-and-competition/. [Accessed 29 August 2017].

Fears, D. (2016). *'Iowa farmers ripped out prairie; now some hope it can save them'*, The Washington Post, 7 August. [Online]. https://www.washingtonpost.com/national/health-science/iowa-farmers-ripped-out-prairie-now-some-hope-it-can-save-them/2016/08/07/1ff747a2-5274-11e6-88eb-7dda4e2f2aec_story.html?tid=a_inl&utm_term=.ae1567aaee4c. [Accessed 29 August 2017].

Federici, S. (2012). *Revolution at Point Zero: Housework, Reproduction, and Feminist Struggle*, PM Press, Oakland, California.

Feldman, J.M. (1991). 'Constituencies and New Markets for Economic Conversion: Reconstructing the United States Physical, Environmental and Social Infrastructure', in Bischak, G. A. (ed.), *Towards a Peace Economy in the United States: Essays on Military Industry, Disarmament and Economic Conversion*, St. Martin's Press, New York, pp. 202–241.

Feldman, J. M. (2006). 'Industrial Conversion: A Linchpin for Disarmament and Development', in Geeraerts, G., Pauwels, N. and Remacle, E. (eds), *Dimensions of Peace and Security: A Reader*, Peter Lang, Brussels, pp. 193–220.

Feldman, J. M. (2007). 'From Warfare State to "Shadow State". Militarism, Economic Depletion, and Reconstruction. From the permanent war economy to perpetual interventionism', *Social Text,* 91, 25(2), pp. 143–168.

Feldman, J. M. (2010). 'The foundations for extending green jobs. The case of the rail-based mass transit sector in North America', *International Journal of Labour Research,* 2(2), pp. 269–291.

Fell, H-J. (2017). 'Accelerating the global transition to 100% renewable energy', The Beam, 4 September. [Online]. Available at:https://medium.com/thebeammagazine/accelerating-the-global-transition-to-100-renewable-energy-by-hans-josef-fell-62889ac8df2e. [Accessed 24 January 2018].

Ferguson, R. S. and Lovell, S. T. (2014). 'Permaculture for agroecology: design, movement, practice, and worldview. A review', *Agronomy for Sustainable Development*, 34(2), pp. 251–274.

Fernandez-Cornejo, J., Nehring, R., Osteen, C., *et al.* (2014). *'Pesticide Use in U.S. Agriculture: 21 Selected Crops, 1960–2008'*, U.S. Department of Agriculture, Economic Research Service, May. [Online]. Available at: https://www.ers.usda.gov/webdocs/publications/43854/46734_eib124.pdf?v=41830. [Accessed 29 August 2017].

Ferris, D. (2015). *'Cuba wants clean energy. Can the U.S. deliver?'*, E&E News, 25 June. [Online]. Available at: http://www.eenews.net/stories/1060020853. [Accessed 29 August 2017].

Fields, A. B. (2003). *Rethinking Human Rights for the New Millennium*, Palgrave MacMillan, New York.

Figueres, C., Schellnhuber, H.J., Whiteman, G., *et al.* (2017). 'Three years to safeguard our climate', *Nature,* 546, pp. 593–595.

Fitz, D. (2014). '"How Green is the 'Green New Deal'?', *Links International Journal of Socialist Renewal*, 9 July. [Online]. Available at: http://links.org.au/node/3943. [Accessed 29 August 2017].

Fitzgerald, H. (2015) *'Costco could beat Whole Foods as the nation's top seller of organic food'*, Business Insider, 5 June. [Online]. Available at: http://www.businessinsider.com/costco-becomes-top-seller-of-organic-food-2015-6. [Accessed 26 January 2018].

FitzRoy, F. (2017). 'How the renewable energy transition could usher in an economic revolution', Symposium: Pathways to the Post-Carbon Economy, 27 July. [Online]. Available at: https://medium.com/insurge-intelligence/the-costs-of-transition-to-renewable-energy-and-a-low-carbon-economy-161026345583. [Accessed 25 January 2018].

Foley, J. A., Ramankutty, N., Brauman, K. A., *et al.* (2011). 'Solutions for a cultivated planet', *Nature,* 478, pp. 337–342.

Ford, K. W., Rocklin, G. I., and Scolow, R. H. (eds). (1975). *Efficient Use of Energy, Part I — A Physics Perspective,* American Institute of Physics, New York.

Forsythe, M. (2017). *'China Aims to Spend at Least $360 Billion on Renewable Energy by 2020'*, New York Times, 5 January. [Online]. Available at: https://www.nytimes.com/2017/01/05/world/asia/china-renewable-energy-investment.html. [Accessed 29 August 2017].

Fortuna, C. (2017). *'Divestment Year in Review 2017*'. CleanTechnica, 23 December, [Online]. Available at: https://cleantechnica.com/2017/12/23/cleantechnica-divestment-year-review-2017/, [Accessed24 January 2018].

Foster, J. B. and Burkett, P. (2008). 'Classical Marxism and the Second Law of Thermodynamics: Marx/Engels, the Heat Death of the Universe Hypothesis, and the Origins of Ecological Economics', *Organization & Environment,* 21(1), pp. 3–37.

Foster, J. B. and Burkett, P. (2016). *Marx and the Earth: An Anti-Critique,* Brill, Leiden.

Foster, J. B. and Magdoff, F. (1998). 'Liebig, Marx and the Depletion of the Natural Fertility of the Soil: Implications for Sustainable Agriculture', *Monthly Review*, 50(3), pp. 32–45.

Foster, J. B., Clark, B. and York, R. (2010). 'Capitalism and the curse of energy efficiency', *Monthly Review,* 62(6), pp. 1–12.

Frankfurt School. (2016). *'Global Trends in Renewable Energy'*, Frankfurt School of Finance & Management. [Online]. Available at: http://fs-unep-centre.org/sites/default/files/publications/globaltrendsinrenewableenergyinvestment-2016lowres_0.pdf. [Accessed 23 August 2017].

Frase, P. (2015). *'Ours to Master'*, Jacobin, 18 March, [Online]. Available at: https://www.jacobinmag.com/2015/03/automation-frase-robots/. [Accessed 29 August 2017].

Frase, P. (2016). *Four Futures. Visions of the World After Capitalism,* Verso, London.

Frazier, I. (2017). *'The Vertical Farm'*, The New Yorker, 9 January. [Online]. Available at: http://www.newyorker.com/magazine/2017/01/09/the-vertical-farm. [Accessed 29 August 2017].

Frolov, I. T. (1984). *Dictionary of Philosophy*, International Publishers, New York.

Fuss, S. (2016). 'Substantial risk for financial assets', *Nature Climate Change,* 6, pp. 659–660.

Fustier, K., Gray, G., Gundersen, C., *et al.* (2016). *'Global oil supply: Will mature field declines drive the next supply crunch?'*, HSBC. [Online]. Available at: https://drive.google.com/file/d/0B9wSgViWVAfzUEgzMlBfR3UxNDg/view. [Accessed 10 February 2017].

Garcia, M. A. and Altieri, M. A. (2005). 'Transgenic Crops: Implications for Biodiversity and Sustainable Agriculture', *Bulletin of Science, Technology & Society*, 25(4), August, pp. 335–353. DOI: 10.1177/0270467605277293.

Gardner, G. (2013). 'Conserving Nonrenewable Resources', in Worldwatch Institute (eds.), *State of the World 2013*, Island Press, Washington, DC, pp. 99–109.

Garrett, L. (2015). *'Global Health Goal Hits and Misses'*, 7 April. [Online]. Available at: http://www.cfr.org/development/global-health-goal-hits-misses/p36405. [Accessed 23 August 2017].

Garzon, P. and Salazar-Lopez, L. (2017). *'China's Other Big Export: Pollution'*, New York Times, 21 July. [Online]. Available at: https://www.nytimes.

com/2017/07/21/opinion/china-climate-pollution-global-warming.html. [Accessed 29 August 2017].

Gasser, T., Guivarch, C., Tachiiri, K., *et al.* (2015). 'Negative emissions physically needed to keep global warming below 2 °C', *Nature Communications,* 6, p. 7958.

Georgescu-Roegen, N. (1971). *The Entropy Law and the Economic Process*, Harvard University Press, Cambridge.

Georgescu-Roegen, N. (1976). *Energy and Economic Myths*, Pergamon Press, New York.

Georgescu-Roegen, N. (1981). 'Energy, matter, and economic valuation: Where do we stand?', in Daly, H. E. and Umana, A. F. (eds.), *Energy, Economics, and the Environment, AAAS Selected Symposium 64*, Boulder, Westview Press, Colorado, pp. 43–79.

Georgescu-Roegen, N. (1986). 'The Entropy Law and the Economic Process in Retrospect', *Eastern Economic Journal*, 12(1), pp.3-25.

Georgescu-Roegen, N. (1989). 'Afterword', in Rifkin, J., *Entropy*. Revised edition Bantam Books, New York, pp. 261–269.

Geuss, M. (2017). *'The world spent less money to add more renewable than ever in 2016'*, Arstechnica, 10 April. [Online]. Available at: https://arstechnica.com/information-technology/2017/04/global-investment-in-renewables-fell-in-2016-but-thats-not-a-bad-sign/. [Accessed 23 August 2017].

Gilabert, P. (2015). 'The Socialist Principle "From Each According To Their Abilities, To Each According To Their Needs', *Journal of Social Philosophy*, 46(2), pp. 197–225.

Gindin, S. (2016). *'Chasing Utopia, Worker Ownership and Cooperatives Will Not Succeed by Competing on Capitalism's Terms'*, Jacobin, 10 March. [Online]. Available at: https://www.jacobinmag.com/2016/03/workers-control-coops-wright-wolff-alperovitz/. [Accessed 29 August 2017].

Ginsberg, A. *(1956). Howl and Other Poems,* City Lights Books, San Francisco.

Global Agriculture. (2017). 'Wastewater is an untapped resource for agriculture, says World Water Report'. [Online]. Available at: http://www.globalagriculture.org/whats-new/news/en/32496.html. [Accessed 24 August 2017].

Global Wind Energy Council. (2017). *'Global Statistics'*. [Online]. Available at: http://www.gwec.net/global-figures/graphs/. [Accessed 18 August 2017].

Glover, J. D., Cox, C. M., and Reganold, J. P. (2007). 'Future Farming: A Return to Roots', *Scientific American*, August, pp. 82–89.

Godfray, H., Charles, J., Beddington, J. R., *et al.* (2010). 'Food Security: The Challenge of Feeding 9 Billion People', *Science*, 327(5967), pp. 812–818.

Goldblatt, C. and Watson, A. J. (2012). 'The Runaway Greenhouse: Implications for Future Climate Change, Geoengineering and Planetary Atmospheres', *Philosophical Transactions of the Royal Society A*, 370, pp. 4197–4216.

Goodwin, P., Katvouta, A., Roussenov, V. M., *et al.* (2018). 'Pathways to 1.5 °C and 2 °C warming based on observational and geological constraints', *Nature Geoscience*, DOI: 10.1038/s41561-017-0054-8.

Gordon, T. and Webber, J. R. (2016). *Blood of Extraction: Canadian Imperialism in Latin America,* Fernwood, Nova Scotia.

Gorz, A. (1967). *Strategy for Labor*, Beacon Press, Boston.

Gould, S. and Bender, J. (2015). *'These charts show the immensity of the US' defense budget'*, Business Insider Singapore, 31 August. [Online]. Available at: http://www.businessinsider.com/the-us-defense-budget-is-massive-2015-8. [Accessed 29 August 2017].

Goulson, D., Nicholls, E., Botias, C., *et al.* (2015). 'Bee declines driven by combined stress from parasites, pesticides, and lack of flowers', *Science*, 347(6229), 1255957, DOI: 10.1126/science.1255957.

Gramsci, A. (2011). *Letters from Prison*, Columbia University Press, New York.

Green Party of the United States. (2016). *'The Green New Deal'*, [Online]. Available at: http://gpus.org/organizing-tools/the-green-new-deal/. [Accessed 23 August 2017].

Greenbaum, S. (2014). *'D-Town Farm: A Modern- Day Manifestation of an Ongoing Historical Battle for Equality'*. [Online]. Available at: https://detroitenvironment.lsa.umich.edu/d-town-farm-a-modern-day-manifestation-of-an-ongoing-historical-battle-for-equality/. [Accessed 29 August 2017].

Griffis, T. J., Chen, Z., Baker, J. M., *et al.* (2017). 'Nitrous oxide emissions are enhanced in a warmer and wetter world', *Proceedings of the National Academy of Sciences (USA)*, 114(45), pp. 12081–12085.

Griffiths, J. (2016). *'Millions of Chinese workers go unpaid amid economic slowdown'* CNN, [Online]. (Updated 1 May 2016). Available at: http://www.cnn.com/2016/04/30/asia/china-on-strike-worker-protests/index.html. [Accessed 29 August 2017].

Griscoma, B. W., Adams. J., Ellis, P. W., *et al.* (2017). 'Natural climate solutions', *Proceedings of the National Academy of Sciences (USA)*, 114(44), pp. 11645–11650.

Gronlund, L. (2013). *'How Much Does it Cost to Create a Single Nuclear Weapon?'*, Union of Concerned Scientists, November. [Online]. Available at: http://www.ucsusa.org/publications/ask/2013/nuclear-weapon-cost.html. [Accessed 29 August 2017].

Grube, A., Donaldson, D., Kiely, T., *et al.* (2011). *'Pesticides industry sales and usage: 2006 and 2007 market estimates'*, US Environmental Protection Agency. [Online]. Available at: www.epa.gov/opp00001/pestsales/07pestsales/market_estimates2007.pdf. [Accessed 29 August 2017].

Gunaratna, S. (2017). *'Earth faces 'biological annihilation' in sixth mass extinction, scientists warn'*, CBS News, 10 July. [Online]. Available at: http://www.cbsnews.com/news/sixth-mass-extinction-biological-annihilation/. [Accessed 29 August 2017].

Gurian-Sherman, D. (2009). *'Failure to Yield: Evaluating the Performance of Genetically Engineered Crops'*, Union of Concerned Scientists, April. [Online]. Available at: http://www.ucsusa.org/sites/default/files/legacy/assets/documents/food_and_agriculture/failure-to-yield.pdf. [Accessed 29 August 2017].

Haas, B. (2016). *'China riot police seal off city centre after smog protesters put masks on statues'*, The Guardian, 12 December. [Online]. Available at: https://www.theguardian.com/world/2016/dec/12/china-riot-police-seal-off-city-centre-after-smog-protesters-put-masks-on-statues. [Accessed 29 August 2017].

Haeckel, E. (1900). *The Riddle of the Universe,* Harper & Row, New York.

Hakim, D. (2016). *'Doubts About the Promised Bounty of Genetically Modified Crops',* New York Times, 30 October. [Online]. Available at: https://www.nytimes.com/2016/10/30/business/gmo-promise-falls-short.html. [Accessed 29 August 2017].

Hall, C. A. S. Lambert, J. G., and Balogh. S. B. (2014). 'EROI of different fuels and the implications for society', *Energy Policy,* 64, pp. 141-152.

Hallen, P. (1987). *'Making Peace With Nature: Why Ecology Needs Feminism'*, The Trumpter, 4(3). [Online]. PDF Available at: http://trumpeter.athabascau.ca/index.php/trumpet/article/view/613. [Accessed 29 August 2017].

Halweil, B. (2006). *'Can Organic Farming Feed U.S. All?'*, World Watch Institute, 19(3). [Online]. Available at: http://www.worldwatch.org/node/4060. [Accessed 29 August 2017].

Hamouchene, H. (2015). *'Desertec: the renewable energy grab?'*, 1 March. [Online]. Available at: https://newint.org/features/2015/03/01/desertec-long. [Accessed 24 August 2017].

Hamouchene, H. (2016). *'The Ouarzazate Solar Plant in Morocco: Triumphal 'Green' Capitalism and the Privatization of Nature'*. [Online]. Available at: http://www.jadaliyya.com/Details/33115/The-Ouarzazate-Solar-Plant-in-Morocco-Triumphal-%60Green%60-Capitalism-and-the-Privatization-of-Nature. [Accessed 26 January 2018].

Hanley, S. (2017). *'Germany Breaks A Solar Record — Gets 85% Of Electricity From Renewables'*, Clean Technica, 8 May. [Online]. Available at: https://cleantechnica.com/2017/05/08/germany-breaks-solar-record-gets-85-electricity-renewables. [Accessed 29 August 2017].

Hanley, S. (2018). *'Water-Based Air Conditioning Slashes Energy Usage & Uses No Refrigerants'*, Clean Technica, 10 January. [Online]. Available at: https://cleantechnica.com/2018/01/10/water-based-air-conditioning-slashes-energy-usage-uses-no-refrigerants/. [Accessed 24 January 2018].

Hansen, J. (2013). *'Exaggeration, Jumping the Gun, and Venus Syndrome: Essay on climate science communication'*, 15 April. [Online]. Available at: http://www.columbia.edu/~jeh1/mailings/2013/20130415_Exaggerations.pdf. [Accessed 26 January 2018].

Hansen, J. (2017). '*Young People's Burden: Requirement of Negative CO2 Emissions*', 18 July. [Online]. Available at: http://www.columbia.edu/~jeh1/mailings/2017/20170718_BurdenCommunication.pdf. [Accessed 26 January 2018].

Hansen, J., Kharecha, P., Sato, M., *et al.* (2011). *'The Case for Young People and Nature: A Path to a Healthy, Natural, Prosperous Future'*. [Online]. Available at: http://www.climatecrime.net/uploads/_Hansen_Case_For_Young_People.pdf. [Accessed 26 January 2018].

Hansen, J., Kharecha, P., Sato, M., *et al.* (2013). 'Assessing 'Dangerous Climate Change': Required Reduction of Carbon Emissions to Protect Young People', Future Generations and Nature, *PLOS ONE,* 8(12), e81648.

Hansen, J., Sato, M., Hearty, P., *et al.* (2016).' Ice melt, sea level rise and superstorms: evidence from paleoclimate data, climate modeling, and modern observations that 2° C global warming could be dangerous', *Atmos. Chem. Phys.,* 16, pp. 3761–3812.

Hansen, J., Sato, M., Kharecha, P., *et al.* (2017). 'Young people's burden: requirement of negative CO_2 Emissions, *Earth Syst. Dynam.*, 8, pp. 577–616.

Hardin, G. (1968). 'The Tragedy of the Commons', *Science*, 162(3859), pp. 1243–1248.

Harrabin, R. (2017a). *'Huge plastic waste footprint revealed'*, BBC, 15 March. [Online]. Available at: http://www.bbc.com/news/science-environment-39279392. [Accessed 29 August 2017].

Harrabin, R. (2017b). *'World's first floating wind farm emerges off coast of Scotland'*, BBC, 23 July. [Online]. Available at: http://www.bbc.com/news/business-40699979. [Accessed 29 August 2017].

Harris, J. (2005). 'To Be or Not To Be: The Nation-Centric World Order Under Globalization', *Science & Society*, 69(3), pp. 329–340.

Harvey, D. (2010). *The Enigma of Capital: and the Crises of Capitalism (2nd Edition),* Oxford University Press, New York.

Harvey, F. (2011). *'World Headed for Irreversible Climate Change in Five Years, IEA warns'*, The Guardian, 9 November. [Online]. Available at: https://www.theguardian.com/environment/2011/nov/09/fossil-fuel-infrastructure-climate-change.[Accessed 29 August 2017].

Harvey, R. (2016). *'NIMBYism, co-operatives and Germany's Energy Transition'*, The News 12 September. [Online]. Available at: https://www.thenews.coop/108839/sector/energy/nimbyism-co-operatives-germanys-energy-transition. [Accessed 29 August 2017].

Hass, A. (2016). *'Israel Incapable of Telling Truth About Water It Steals From Palestinians'*, Haaretz, 22 June. [Online]. Available at: http://www.haaretz.com/opinion/.premium-1.726350. [Accessed 29 August 2017].

Hawken, P. (2007). *Blessed Unrest: How the Largest Movement in the World Came into Being and Why No One Saw It Coming,* Penguin, New York.

He, Y., Trumbore, S. E., Torn, M. S., *et al.* (2016). 'Radiocarbon constraints imply reduced carbon uptake by soils during the 21st Century', *Science,* 353(6306), pp. 1419–1424.

Healthy Living (2013). *'U.S. Life Expectancy Ranks 26th In The World, OECD Report Shows'*, The Huffington Post, 21 November. [Online]. Available at: http://www.huffingtonpost.com/2013/11/21/us-life-expectancy-oecd_n_4317367.html?view=print&comm_ref=false. [Accessed 23 August 2017].

Heinberg, R. (2009). *Searching for a miracle: "Net energy" limits & the fate of industrial society*, International Forum on Globalization and the Post Carbon Institute.

Heinberg, R. (2016). *'Is the Oil Industry Dying?'*, PS Mag, 10 August, [Online]. Available at: https://psmag.com/news/is-the-oil-industry-dying#.p5kh8w3qk. [Accessed 23 August 2017].

Heinrich, M. (2004). *An Introduction to the Three Volumes of Karl Marx's Capital,* Monthly Review Press, New York.

Heinrich, M. (2013). 'The "Fragment on Machines": The Grundrisse and its Overcoming in Capital', in Bellofiore, R., Starosta, G. and Thomas, P. D. (eds.), *Marx's Laboratory*, Haymarket, Chicago, pp. 197–212.

Hendrick, M. F., Cleveland, S. and Phillips, N. G. (2016). 'Unleakable carbon', *Climate Policy*, 17(8), pp. 1057–1064.

Henley, B. J. and King, A. D. (2017). 'Trajectories towards the 1.5° Paris target: Modulation by the Interdecadal Pacific Oscillation', *Geophysical Research Letters,* 44(9), pp. 4256–4262.

Hertsgaard, M. (2012). *'Climate Change Kills 400,000 a Year, New Report Reveals'*, The Daily Beast, 27 September. [Online]. Available at: http://www.thedailybeast.com/climate-change-kills-400000-a-year-new-report-reveals. [Accessed 23 August 2017].

Hill, J. S. (2017). *'100% Global Renewable Electricity No Longer Flight Of Fancy, More Cost-Effective Than Current System*', Clean Technica, 9 November. [Online]. Available at: https://cleantechnica.com/2017/11/09/100-global-renewable-electricity-no-longer-flight-fancy-cost-effective-current-system/. [Accessed 24 January 2018].

Ho, M. W. and Ulanowicz, R. (2005). 'Sustainable systems as organisms?', *BioSystems*, 82, pp. 39–51.

Hoekstra, A. Y. (2008). *'The water footprint of food'*, Water Footprint. [Online]. Available at: http://waterfootprint.org/media/downloads/Hoekstra-2008-WaterfootprintFood.pdf. [Accessed 29 August 2017].

Holleman, H. (2017). '*Capital* and ecology', in: Schmidt, I. and Fanelli, C. (eds.), *Reading 'Capital' Today*, Pluto Press, London, pp. 160–180.

Hollender, R. (2015). 'Post-growth in the Global South: The Emergence of Alternatives to Development in Latin America', *Socialism and Democracy,* 29(1), pp. 73–101.

Holmgren, D. (2002). *Permaculture: Principles & Pathways Beyond Sustainability*, Holmgren Design Services.

Holt-Gimenez, E. (2007). *'The biofuel myths'*, New York Times, 10 July. [Online]. Available at: http://www.nytimes.com/2007/07/10/opinion/10iht-edholt.1.6588231.html. [Accessed 29 August 2017].

Hong, W. L., Torres, M. E., Carrol, J., *et al*. (2017). 'Seepage from an arctic shallow marine gas hydrate reservoir is insensitive to momentary ocean warming', *Nature Communications*, 8, 15745, DOI: 10.1038/ncomms 15745.

House, K. Z., Baclig, A. C., Ranjan, M., *et al.* (2011). 'Economic and energetic analysis of capturing CO_2 from ambient air', *Proceedings of the National Academy of Sciences (USA),* 108 (51), pp. 20428–20433.

Howard, P. (2016). *'Organic Chart Infographic'.* [Online]. Available at: https://www.cornucopia.org/wp-content/uploads/2016/01/Organic-chart-Jan-2016.jpg. [Accessed 29 August 2017].

Howarth, R. W. (2014). 'A Bridge to Nowhere: Methane Emissions and the Greenhouse Gas Footprint of Natural Gas', *Energy Science & Engineering,* 2(2), pp. 47–60.

Hrala, J. (2016). *'This World-First Farm Grows Vegetables in The Desert With Nothing But Sun and Seawater'*, Science Alert, 7 October. [Online]. Available at: https://www.sciencealert.com/this-farm-uses-sun-and-seawater-to-grow-vegetables-in-the-desert. [Accessed 29 August 2017].

Huber, M. T. (2017). 'Value, Nature, and Labor: A Defense of Marx', *Capitalism Nature Socialism,* 28(1), pp. 39–52.

Hudak, M. (2017). *'Claims that Livestock Grazing Enhances Soil Sequestration of Atmospheric Carbon Are Outweighed by Methane Emissions From Enteric Fermentation: A Closer Look at Franzluebbers and Stuedemann (2009)'*, 7 April 2015. [Online]. Available at: http://www.mikehudak.com/Articles/FranzluebbersAndStuedemannCritique.html. [Accessed 20 August 2017].

Hvistendahl, M. (2012). *Unnatural Selection: Choosing Boys Over Girls, and the Consequences of a World Full of Men,* Public Affairs, New York.

Hvistendahl, M. (2017). 'China aims to sow a revolution with GM seed takeover', *Science,* 356, pp. 16–17.

ICPR. (2016). *'World Prison Brief'*, International for Criminal Policy Research. [Online]. Available at: http://www.prisonstudies.org/highest-to-lowest/prison_population_rate?field_region_taxonomy_tid=All. [Accessed 30 August 2017].

IESA. (2016). 'Venezuela Energy in Figures 2014–2015', Instituto de Estudios Superiores de Administración, Venezuela. [Online]. Available at: http://servicios.iesa.edu.ve/portal/ciea//eec%202014%20iesa%20ingles.pdf. [Accessed 26 January 2018].

Indigenous Environmental Network (2016). '*Carbon Offsets cause Conflict and Colonialism, Indigenous Peoples denounce at United Nations; Demand Cancellation of REDD+*', 18 May. [Online]. Available at: http://www.ienearth.

org/carbon-offsets-cause-conflict-and-colonialism/. [Accessed 27 January 2018].

International Energy Outlook (2016). '*International Energy Outlook (2016)*', U.S. Energy Information Administration. Washington, DC. [Online]. Available at: https://www.eia.gov/outlooks/ieo/pdf/0484(2016).pdf. [Accessed 27 January 2018].

International Organization of Motor Vehicle Manufacturers (2014). *'2014 Production Statistics'*. [Online]. Available at: http://www.oica.net/category/production-statistics/. [Accessed 29 August 2017].

International Trade Union Confederation. (2011). *'Unions welcome pathway to green economy in UN environment report'*, Ecosocialism Canada blog, 18 March. [Online]. Available at: http://ecosocialismcanada.blogspot.com/2011/03/unions-welcome-pathway-to-green-economy.html. [Accessed 29 August 2017].

IPCC. (2007). 'Climate Change 2007: Synthesis Report', in Pachauri, R. K. and Reisinger, A. (eds.), *Contribution of Working Groups I, II and III to the Fourth Assessment Report of the Intergovernmental Panel on Climate Change*, Geneva, Switzerland, pp. 1–112.

IPCC. (2013). 'Summary for Policymakers,' in Stocker, T. F., Qin, D., Plattner, G.-K., *et al.* (eds.), *Climate Change 2013: The Physical Science Basis. Contribution of Working Group I to the Fifth Assessment Report of the Intergovernmental Panel on Climate Change,* Cambridge University Press, Cambridge, United Kingdom and New York, USA, pp. 1–33.

IPCC. (2014). 'Climate Change 2014: Synthesis Report'. *Contribution of Working Groups I, II and III to the Fifth Assessment Report of the Intergovernmental Panel on Climate Change,* Core Writing Team, Pachauri, R. K. and Meyer, L. A. (eds.). IPCC, Geneva, Switzerland, pp. 1–169.

IPU. (2017). *'Women in National Parliaments'*, Inter-Parliamentary Union. (Updated 1 July 2017). [Online]. Available at: http://www.ipu.org/wmn-e/classif.htm. [Accessed 29 August 2017].

IRENA (2015). 'Renewable Power Generation Costs in 2014', [Online]. Available at: http://www.irena.org/media/Files/IRENA/Agency/Publication/2015/IRENA_RE_Power_Costs_2014_report.pdf. [Accessed 26 January 2018].

IRENA (2017). *Synergies between renewable energy and energy efficiency, a working paper based on REmap*, International Renewable Energy Agency, Abu Dhabi.

ITUC. (2011). *'Unions Welcome Pathway to Green Economy in UN Environment Report'*, International Trade Union Confederation, 21 February. [Online]. Available at: https://www.ituc-csi.org/unions-welcome-pathway-to-green,8607?lang=en. [Accessed 29 August 2017].

Jackson, T. (2009). Prosperity Without Growth: Economics for a Finite Planet. Earthscan, London.

Jacobin. (2017). [Online]. Available at: https://www.jacobinmag.com/about. [Accessed 29 August 2017].

Jacobson, M. Z. (2017a). *'Response to Forbes: Stop Inaccuracies—100% Renewable Energy Is Possible'*, Ecowatch, 6 July. [Online]. Available at: https://www.ecowatch.com/mark-jacobson-pnas-2454280135.html. [Accessed 26 January 2018].

Jacobson, M. Z. (2017b). 'Roadmaps to transition countries to 100% clean, renewable energy for all purposes to curtail global warming, Air Pollution, and Energy Risk', *Earth's Future*, 5, pp. 948–952.

Jacobson, M. Z. and Delucchi, M. A. (2009). 'A Path to Sustainable Energy by 2030', *Scientific American* 301, pp. 58–65.

Jacobson, M. Z. and Delucchi, M. A. (2011). 'Providing all global energy with wind, water, and solar power, Part I: Technologies, energy resources, quantities and areas of infrastructure, and materials', *Energy Policy,* 39, pp. 1154–1169.

Jacobson, M. Z., Delucchi, M. A., Ingraffea, A. R., *et al.* (2014). 'A roadmap for repowering California for all purposes with wind, water and sunlight', *Energy,* 73, pp. 875–889.

Jacobson, M. Z., Delucchi, M. A., Bazouin, G., *et al.* (2015). '100% clean and renewable wind, water, and sunlight (WWS) all-sector energy roadmaps for the 50 United States', *Energy Environ. Sci.*, 8, pp. 2093–2117.

Jacobson, M. Z., Delucchi, M. Z., Cameron, M. A., *et al.* (2017a). 'The United States can keep the grid stable at low cost with 100% clean, renewable energy in all sectors despite inaccurate claims', *Proc Natl Acad Sci (USA)*, 114 (26), pp. E5021-E5023.

Jacobson, M. Z., Delucchi, M. Z., Bauer, Z. A. F., *et al.* (2017b). '100% Clean and Renewable Wind, Water, and Sunlight (WWS) All-Sector Energy Roadmaps for 139 Countries of the World', *Joule*, 1, pp. 108–121.

Jakob, M. and Hilaire, J. (2015). 'Climate science: Unburnable fossil-fuel reserves', *Nature,* 517 (7533), January 8, pp. 150–152.

Jamail, D. (2017a). *'Release of Arctic Methane "May Be Apocalyptic," Study Warns'*, Truthout, 23 March [Online]. Available at: http://www.truth-out.org/

news/item/39957-release-of-arctic-methane-may-be-apocalyptic-study-warns. [Accessed 29 August 2017].

Jamail, D. (2017b). *'Scientists Warn of "Biological Annihilation" as Warming Reaches Levels Unseen for 115,000 Years'*, Truthout, 31 July. [Online]. Available at: http://www.truth-out.org/news/item/41425-biological-annihilation-trillion-ton-icebergs-warming-levels-unseen-for-115-000-years. [Accessed 29 August 2017].

James, S. (2012). *Sex, Race, and Class*, PM Press, Oakland, California.

Jenkins, J. (2015). *'How Much Land Does Solar, Wind and Nuclear Energy Require?'*, The Energy Collective, 25 June. [Online]. Available at: http://www.theenergycollective.com/jessejenkins/2242632/how-much-land-does-solar-wind-and-nuclear-energy-require. [Accessed 29 August 2017].

Jenkins, K., McCauley, D., Heffron, R., *et al.* (2016) 'Environmental justice: A conceptual review', *Energy Research & Social Science*, 11, pp. 174–182.

Jensen, D. (2006). *Endgame, Vol. 1: The Problem of Civilization,* Seven Stories Press, New York.

Jensen, D. (2012). 'Self-evident Truths', *Orion,* 31(4), pp. 12–13.

Joensen, L., Semino, S. and Paul, H. (2005) *'Argentina: A Case Study of the Impact of Genetically Engineered Soya'*, The Gaia Foundation, London. [Online]. Available at: http://www.econexus.info/sites/econexus/files/ENx-Argentina-GE-Soya-Report-2005.pdf. [Accessed 24 February 2018].

Johnson, J. (2011). Long History of U.S. Energy Subsidies, *Chemical & Engineering News,* 89(51), pp. 30–31.

Jones, V. (2008). *The Green Collar Economy,* HarperCollins, New York.

Kaberger, T. and Mansson, B. (2001). 'Entropy and Economic Processes – Physics Perspectives', *Ecological Economics*, 36(1), pp. 165–179.

Kallis, G. and Swyngedouw, E. (2017). *'Do Bees Produce Value? A Conversation between an ecological economist and a Marxist geographer'*, Capitalism Nature Socialism. [Online]. Available at: http://www.cnsjournal.org/do-bees-produce-value-a-conversation-between-an- ecological-economist-and-a-marxist-geographer/. [Accessed 21 August 2017].

Kastner R. (2016). 'Hope for the future: how farmers can reverse climate change', *Socialism and Democracy,* 30(2), pp. 154–170.

Kavdir, Y., Zhang, W., Basso, B., *et al.* (2014). 'Development of a new long-term drought resilient soil water retention technology', *Journal of Soil and Water Conservation*, 69, Sept/Oct, pp. 154A-160A.

Kawachi, I. and Kennedy, B. P. (2006). *The Health of Nations: Why Inequality Is Harmful to Your Health,* The New Press, New York.

Kempton,W., Pimenta, F. M., Veron, D. E., *et al.* (2010). 'Electric power from offshore wind via synoptic-scale interconnection', *Proc Natl Acad Sci (USA)*, 107, pp. 7240–7245.

Kids Count (2014). *'Child Poverty by Ward'*, Kids Count Data Center. [Online]. Available at: http://datacenter.kidscount.org/data/Tables/6748-child-poverty-by-ward?loc=10&loct=3#detailed/21/1852-1859/true/36/any/13834. [Accessed 23 August 2017].

Kirschenmann, F. (2004). 'Ecological Morality: A New Ethic for Agriculture', in, Rickerl, D. and Francis, C. (eds.), *Agroecosystems Analysis*, American Society of Agronomy, Madison, WI, pp. 167–176.

Klare, M. T. (2002). *Resource Wars,* Henry Holt & Company, New York.

Klare, M. T. (2007). *'The Pentagon vs. peak oil'*, 15 June 2007.[Online]. Available at: http://www.truthdig.com/report/item/20070615_the_pentagon_v_peak_oil/. [Accessed 29 August 2017].

Klare, M. T. (2015). '*Tomgram: Michael Klare, Are Resource Wars Our Future?*', 3 November. [Online]. Available to: http://www.tomdispatch.com/blog/176063/tomgram%3A_michael_klare,_are_resource_wars_our_future. [Accessed 30 January 2018].

Kleidon, A., Miller, L., and Gans, F. (2015). 'Physical Limits of Solar Energy Conversion in the Earth System', in Tüysüz, H. and Chan, C. K. (eds), *Solar Energy for Fuels*, Springer Nature, Berlin, pp. 1–22.

Klein, N. (2008) *The Shock Doctrine: The Rise of Disaster Capitalism*, Picador, New York.

Klein, N. (2014). *This Changes Everything: Capitalism vs the Climate*, Simon & Schuster, New York.

Knopf , B., Fuss, S., Hansen, G., *et al.* (2017). 'From Targets to Action: Rolling up our Sleeves after Paris', *Global Challengers,* 1(2), 1600007.

Kolbert, E. (2015). *The Sixth Extinction: An Unnatural History,* Henry Holt and Company, New York.

Knuth, S. (2017). 'Green devaluation: disruption, divestment, and decommodification for a green economy', *Capitalism Nature Socialism,* 28(1), pp. 98–117.

Kornei, K. (2017). *'Here are some of the world's worst cities for air quality'*, Science News, 21 March. [Online]. Available at: http://www.sciencemag.org/news/2017/03/here-are-some-world-s-worst-cities-air-quality. [Accessed 29 August 2017].

Kovel, J. (2002). *The Enemy of Nature: The End of Capitalism or the End of the World?,* Zed Books, New York.

Kovel, J. (2017). 'Ernst Bloch as an ecosocialist thinker for the present crisis', Presentation at *Left Forum,* 2 June, John Jay College of Criminal Justice, New York, N.Y.

KPMG. (2012) *'Expect the Unexpected: Building business value in a changing world'*, KPMG International. [Online]. Available at: https://home.kpmg.com/content/dam/kpmg/pdf/2012/08/building-business-value-part-1.pdf. [Accessed 26 January 2018].

Krausmann, F., Wiedenhofer, D., Lauk, C., *et al.* (2017). 'Global socioeconomic material stocks rise 23-fold over the 20th Century and require half of annual resource use', *Proc Natl Acad Sci (USA),* 114(8), pp. 1880–1885.

Kriegler, E., Edenhofer, O., Reuster, L., *et al.* (2013). 'Is atmospheric carbon dioxide removal a game changer for climate change mitigation?', *Climatic Change*, 118, pp. 45–57.

Kristof, N.D. and WuDunn, S. (2010). *Half the Sky: Turning Oppression into Opportunity for Women Worldwide*, Vintage, New York.

Kubiszewski, I., Cleveland, C. J. and Endres, P. K. (2010). 'Meta-analysis of net energy return for wind power systems', *Renewable Energy,* 35, pp. 218–225.

Kunze, C. and Becker, S. (2015). 'Collective ownership in renewable energy and opportunities for sustainable degrowth', *Sustain. Sci.,* 10, pp. 425–437.

Kuznetsov, B. G. (1977). *Philosophy of Optimism*, Progress Publishers, Moscow.

LaFond, K. (2015). *'Infographic: California Freshwater Withdrawals'*, Circle of Blue, 14 May. [Online]. Available at: http://www.circleofblue.org/2015/world/infographic-california-freshwater-withdrawals/. [Accessed 24 August 2017].

Laibman, D. (2012). 'Editorial Perspectives: Whither the Occupy Movement: Models and Proposals', *Science & Society,* 76(3), pp. 283–288.

Laibman, D. (2013). 'Editorial Perspectives: The Human-Nature interface: Navigating between Utopic and Dystopic Determinisms', *Science & Society*, 77(2), pp. 145–150.

Landberg, R. (2015). *'For Nuclear's Cost, U.K. Could Have Six Times the Wind Capacity'*, 21 October. [Online]. Available at: http://www.bloomberg.com/news/articles/2015-10-21/for-nuclear-s-cost-u-k-could-have-six-times-the-wind-capacity. [Accessed 29 August 2017].

The Land Institute. (2017). *'Perennial Crops: New Hardware for Agriculture'*. [Online]. Available at: https://landinstitute.org/our-work/perennial-crops/. [Accessed 29 August 2017].

LaPorte, N. (2013). *'Coffee's Economics, Rewritten by Farmers'*, New York Times, 16 March. [Online]. Available at: http://www.nytimes.com/2013/03/17/business/coffees-economics-rewritten-by-farmers.html?mcubz=3. [Accessed 29 August 2017].

Larsen, S. (2015). *'How Much Water Does It Take to Grow America's Favorite Foods'*, Smithsonsianmag.com, 10 April. [Online]. Available at: http://www.smithsonianmag.com/smart-news/how-much-water-does-it-take-grow-americas-favorite-foods-180954941/. [Accessed 29 August 2017].

Latouche, S. (2009). *Farewell to growth,* Wiley, New York.

Latouche, S. (2010). 'Can the Left escape economism?', *Capitalism Nature Socialism*, 23(1), pp. 74–78.

Leach, G. (1976). *Energy and Food Production*, IPC Science and Technology Press, Guilford.

The Leap Manifesto. (2017). [Online]. Available at: https://leapmanifesto.org/en/the-leap-manifesto/. [Accessed 18 August 2017].

Leary, J. (2017). *One Shot: Trees as Our Last Chance for Survival,* Trees for the Future, Silver Spring Maryland.

Lebowitz, M. A. (2003). *Beyond Capital (2nd edition)*, Palgrave MacMillan, New York.

Lebowitz, M. A. (2010). *The Socialist Alternative. Real Human Development*, Monthly Review Press, New York.

Lebowitz, M. A. (2017). *'If you don't understand the second product, you understand nothing about Marx's Capital'*, 9 January. [Online]. Available at: http://links.org.au/lebowitz-150-years-capital-second-product. [Accessed 29 August 2017].

Lehninger, A. L. (1965). *Bioenergetics*, W. A. Benjamin, New York.

Leibrock, A. (2014). *'Good Growth: Farmers Markets Still On the Rise'*, Sustainable America, 6 August. [Online]. Available at: http://www.sustainableamerica.org/blog/good-growth-farmers-markets-still-on-the-rise/. [Accessed 29 August 2017].

Leifeld, J. and Menichetti, L. (2018). 'The underappreciated potential of peatlands in global climate change mitigation strategies', *Nature Communications*, 9:1071, DOI: 10.1038/s41467-018-03406-6.

Lelieveld, J., Evans, J. S., Fnais, M., *et al.* (2015). 'The contribution of outdoor air pollution sources to premature mortality on a global scale', *Nature,* 525, pp. 367–371.

Lelieveld, J., Proestos, Y., Hadjinkolaou, P., *et al.* (2016). Strongly increasing heat extremes in the Middle East and North Africa (MENA) in the 21st Century, *Climate Change*, 137(1–2), pp. 245–260.

Lenin, V. I. (1929). *What is to be Done?*, International Publishers, New York.

Lenski, S. M., Keoleian, G. A. and Bolon, K. M. (2010). The impact of 'Cash for Clunkers' on greenhouse gas emissions: a life cycle perspective', *Environ. Res. Lett.,* 5(4), 044003.

Lenton, T. M., Pichler, P. P., and Weisz, H. (2016). 'Revolutions in energy input and material cycling in Earth history and human history, *Earth Syst. Dynam.,* 7, pp. 353–370.

Levin, M. J. (2007). *Pesticides: A Toxic Time Bomb in Our Midst*, Praeger, Westport, Connecticut.

Li, M. (2013). 'The 21st Century: Is There an Alternative (To Socialism)?', *Science & Society*, 77(1), pp. 10–43.

Lineweaver, C. H. (2013). 'A simple treatment of complexity: cosmological entropic boundary conditions on increasing complexity', in Lineweaver, C. H., Davies, P. C. and Ruse, M. (eds.), *Complexity and the Arrow of Time*, Cambridge University Press, Cambridge, pp. 42–67.

Liska, A. J., Yang, H., Milner, M., *et al.* (2014). 'Biofuels from crop residue can reduce soil carbon and increase CO_2 emissions', *Nature Climate Change,* 4, pp. 398–401.

Li, Z., Friedman, E. and Ren, H. (eds.) (2016). *China on Strike: Narratives of Workers' Resistance*, Haymarket Books, Chicago.

Liu, C. (2017). 'Severe weather in a warming climate', *Nature,* 544, pp. 422–423.

Loik, M. E., Carter, S. A., Alers, G., *et al.* (2017), 'Wavelength-Selective Solar Photovoltaic Systems: Powering Greenhouses for Plant Growth at the Food-Energy-Water Nexus', *Earth's Future*, 5, pp. 1044–1053.

López-Ráez, J. A., Bouwmeester, H. and Pozo, M. J. (2012). 'A Novel Land-Energy Use Indicator for Energy Crops', *Agroecology and Strategies for Climate Change. Sustainable Agriculture Reviews*, 8, pp. 135–147.

Lorenzo, S. (2016). *'The big challenge for the G20: aligning financial markets to 1.5°C global warming'*, WWF Climate and Energy Blog, 11 April. [Online]. Available at: http://climate-energy.blogs.panda.org/2016/04/11/big-challenge-align-financial-markets-1-5c-global-warming. [Accessed 25 August 2017].

Lott, M. C. (2011). *'10 Calories in, 1 Calorie Out—The Energy We Spend on Food'*, Scientific American Blog, 1 August. [Online]. Available at: https://blogs.scientificamerican.com/plugged-in/10-calories-in-1-calorie-out-the-energy-we-spend-on-food/. [Accessed 29 August 2017].

LPAA. (2015). *'Soils for Food Security and Climate, 4/1000'*, Lima-Paris Action Agenda. [Online]. Available at: http://newsroom.unfccc.int/lpaa/agriculture/join-the-41000-initiative-soils-for-food-security-and-climate/. [Accessed 20 August 2017].

Lu, X., McElroy, M. B. and Kiviluoma, J. (2009). 'Global Potential for Wind-generated Electricity', *Proceedings of the National Academy of Sciences (U.S.A.),* 106(27), pp. 10933–10938.

Lusk, J. (2016) *'Why Industrial Farms are Good for the Environment'*, New York Times, 23 September. [Online]. Available at: https://www.nytimes.com/2016/09/25/opinion/sunday/why-industrial-farms-are-good-for-the-environment.html. [Accessed 29 August 2017].

Lynas, M. (2016). *'GMO Safety Debate is Over'*, Cornell Alliance for Science, 23 May. [Online]. Available at: http://allianceforscience.cornell.edu/blog/mark-lynas/gmo-safety-debate-over. [Accessed 29 August 2017].

Ma, J. and Xu, J. (2017). 'China's energy rush harming ecosystem', *Nature,* 541, p. 30.

Macdonald, F. (2015). *'This Indoor Farm is 100 Times More Productive Than Outdoor Fields'*, ScienceAlert, 13 January. [Online]. Available at: https://www.sciencealert.com/this-indoor-farm-is-100-times-more-productive-than-an-outdoor-one. [Accessed 29 August 2017].

Macias-Fauria, M. (2018). 'Satellite images show China growing green', Nature, 553, 411–413.

Mahan, S. (2015). *'Study Proves Fossil Fuels Way Worse for Land Use than Renewables'*, Clean Techies, 6 July. [Online]. Available at: http://cleantechies.com/2015/07/06/study-proves-fossil-fuels-way-worse-for-land-use-than-renewables/. [Accessed 29 August 2017].

Makhijani, S. and Ochs, A. (2013). 'Renewable Energy's Natural Resource Impacts', in Worldwatch Institute (ed.), *State of the World 2013*, Island Press, Washington, DC, pp. 84–98.

Malm, A. (2016). *Fossil Capital: The Rise of Steam Power and the Roots of Global Warming*, Verso, London.

Malm, A. (2017). 'Overcoming Contradictions of Climate Change in Short Time We Have', Popular Resistance, 29 December. [Online]. Available at: https://popularresistance.org/overcoming-contradictions-of-climate-change-in-short-time-we-have. [Accessed 24 January 2018].

Malm, A. and Hornborg, A. (2014). 'The geology of mankind? A critique of the Anthropocene narrative', *The Anthropocene Review*, 1(1), pp. 62–69.

Mann, C. C. (2018). *'Can Planet Earth Feed 10 Billion People?'* The Atlantic, 23 January. [Online]. Available at: https://www.msn.com/en-us/news/world/can-planet-earth-feed-10-billion-people/ar-AAv3NPd?li=AA4Zpp&ocid=ientp, [Accessed 24 January 2018].

Mann, M. M. (2017). *'It's a fact: climate change made Hurricane Harvey more deadly'*, The Guardian, 28 August. [Online]. Available at: https://www.theguardian.com/commentisfree/2017/aug/28/climate-change-hurricane-harvey-more-deadly. [Accessed 30 August 2017].

Margolis, J. (2015). *'Growing food in the desert: is this the solution to the world's food crisis?'* The Guardian, 24 November. [Online]. Available at: https://www.theguardian.com/environment/2012/nov/24/growing-food-in-the-desert-crisis. [Accessed 29 August 2017].

Marshall, E. and Aillery, M. (2015). *'Climate Change, Water Scarcity, and Adaptation'*, United States Department of Agriculture, 25 November. [Online]. Available at: https://www.ers.usda.gov/amber-waves/2015/november/climate-change-water-scarcity-and-adaptation/. [Accessed 29 August 2017].

Martinez, R. (2017). *'Extreme Drought in Cuba, Many Are Unaware'*, Havana Times, 25 March. [Online]. Available at: http://www.havanatimes.org/?p=124357. [Accessed 29 August 2017].

Martinez, Y. (2017). *'Cuban Agriculture Moves Toward Ecological Practices'*, Granma, 7 June. [Online]. Available at: http://en.granma.cu/cuba/2017-06-07/cuban-agriculture-moves-toward-ecological-practices. [Accessed 29 August 2017].

Martinez-Alier, J. (1987). *Ecological Economics*, Basil Blackwell, Cambridge.

Martinez-Alier, J. (2012). 'Environmental Justice and Economic Degrowth: An Alliance between Two Movements', *Capitalism Nature Socialism*, 23(1), pp. 51–73.

Marvel, K., Kravitz, B. and Caldeira, K. (2012). 'Geophysical limits to global wind power', *Nature Climate Change,* 3, pp. 118–121.

Marx, K. (1938). *Critique of the Gotha Program*, International Publishers, New York. (Original work published 1891).

Marx, K. (1967). *Capital (Three volumes)*, International Publishers, New York. (Original work published 1875).

Marzec, R. P. (2015). *Militarizing the Environment: Climate Change and the Security State*, University of Minnesota Press, Minneapolis.

Masnadi, M. S. and Brandt, A. R. (2017). 'Climate impacts of oil extraction increase significantly with oilfield age', *Nature Climate Change* 7, pp. 551–556.

Mason, P. (2015). *Postcapitalism: A Guide to Our Future,* Allen Lane, London.

Matthews, H. D. and Caldeira K. (2008). 'Stabilizing climate requires near-zero emissions', *Geophysical Research Letters,* 35(4), DOI: 10.1029/2007GL032388.

Matthews, T. K. R., Wilby, R. L. and Murphy, C. (2017). 'Communicating the deadly consequences of global warming for human heat stress', *Proceedings of the National Academy of Sciences (U.S.A.),* 114(15), pp. 3861–3866.

Mauritsen, T. and Pincus, R. (2017).' Committed warming inferred from observations', *Nature Climate Change,* 7, pp. 652–655.

Mawji, O. (2016). *'Canada's Oil Exports Are Dead Without U.S. Shale Production'*, OilPrice.com, 14 December. [Online]. Available at: http://oilprice.com/Energy/Crude-Oil/Canadas-Oil-Exports-Are-Dead-Without-US-Shale-Production.html. [Accessed 29 August 2017].

McCollum, D., Gould, G. and Greene, D. (2009). *'Greenhouse Gas Emissions From Aviation and Marine Transportation: Mitigation Potential and Policies',* Pew Center on Global Climate Change, December. [Online]. Available at: http://cta.ornl.gov/cta/Publications/Reports/GHG_from_Aviation-Dec2009.pdf. [Accessed 26 January 2018].

McGlade, C. and Ekins, P. (2015). 'The geographical distribution of fossil fuels unused when limiting global warming to 2 °C', *Nature,* 517, pp. 187–190.

McGrath, M. (2016). *'California methane leak "largest in U.S. history"'*, BBC News, 26 February. [Online]. Available at: http://www.bbc.com/news/science-environment-35659947. [Accessed 29 August 2017].

McKibben, B. (2016). *'A World at War'*, New Republic, 16 August. [Online]. Available at: https://newrepublic.com/article/135684/declare-war-climate-change-mobilize-wwii. [Accessed 29 August 2017].

McKibben, B. (2017). *'Keep It 100'*, In These Times, 22 August. [Online]. Available at: http://inthesetimes.com/features/bill_mckibben_renewable_energy_100_percent_solution.html. [Accessed 29 August 2017].

Meadows, D. H., Meadows, D. L., Randers, J., *et al.* (1972). *The Limits to Growth*, Universe Books.

Meadows, D., Randers, J., and Meadows, D. (2002). *'A Synopsis: Limits to Growth: The 30-Year Update'*. [Online]. Available at: http://donellameadows.org/archives/a-synopsis-limits-to-growth-the-30-year-update/. [Accessed 29 August 2017].

Mekonnen, M. M. and Hoekstra, A. Y. (2010a). *'The Green, Blue and Grey Water Footprint of Crops and Derived Crop Products'*, Value of Water Research Report Series No. 47, UNESCO-IHE Institute for Water Education, The Netherlands. [Online]. Available at: http://wfn.project-platforms.com/Reports/Report47-WaterFootprintCrops-Vol1.pdf. [Accessed 24 August 2017].

Mekonnen, M. M. and Hoekstra, A. Y. (2010b). *'The Green, Blue and Grey Water Footprint of Farm Animals and Animal Products,'* Value of Water Research Report Series No. 48, UNESCO-IHE Institute for Water Education, The Netherlands. [Online]. Available at: http://waterfootprint.org/media/downloads/Report-48-WaterFootprint-AnimalProducts-Vol1_1.pdf. [Accessed 24 August 2017].

Melillo, J. M., Frey, S. D., DeAngelis, K. M., *et al.* (2017). 'Long-term pattern and magnitude of soil carbon feedback to the climate system in a warming world', *Science,* 358(6359), pp. 101–105.

Menzel, P. and D'Aluisio, F. (1998). *Man Eating Bugs,* Ten Speed Press, Material World Books, California.

Metcalfe, D. B. (2017). 'Microbial change in warming soils', *Science*, 358(6359), pp. 41–42.

Meyer, R., Cullen, B. R. and Eckard, R. J. (2016). 'Modelling the influence of soil carbon on net greenhouse gas emissions from grazed pastures', *Animal Production Science*, 56(3), pp. 585–593.

Millar, R. J., Fuglestvedt, J. S., Friedlingstein, P., *et al.* (2017). 'Emission budgets and pathways consistent with limiting warming to 1.5 C', *Nature Geoscience,* 10, pp. 741–747.

Miller, G. T., Jr. (1999) *Environmental Science: Working with the Earth*, Wadsworth Publishing Company, Belmont, CA.

Milman, O. (2015). *'James Hansen, father of climate change awareness, calls Paris talks "a fraud"'*, The Guardian, 12 December. [Online]. Available at: http://www.theguardian.com/environment/2015/dec/12/james-hansen-climate-change-paris-talks-fraud. [Accessed 29 August 2017].

Milman, O. (2017). *'Meat Industry Blamed for Largest-Ever 'Dead Zone' in Gulf of Mexico'*, The Guardian, 1 August. [Online]. Available at: https://www.theguardian.com/environment/2017/aug/01/meat-industry-dead-zone-gulf-of-mexico-environment-pollution. [Accessed 26 January 2018].

Milton, J. (1667, reprinted 2011). *Paradise Lost*, Penguin Classics, London.

Miroff, N. (2015). *'An arugula-growing farmer feeds a culinary revolution in Cuba'*, The Washington Post, 21 August. [Online]. Available at: http://www.ticotimes.net/2015/08/22/an-arugula-growing-farmer-feeds-a-culinary-revolution-in-cuba. [Accessed 26 January 2018].

Moodliar, M. (2015). 'System Change Without Class Struggle?', *Socialism and Democracy,* 29(1), pp. 141–151.

Mooney, C. (2017a). *'We only have a 5 percent chance of avoiding "dangerous" global warming, a study finds'*, The Washington Post, 31 July. [Online]. Available at: https://www.washingtonpost.com/news/energy-environment/wp/2017/07/31/we-only-have-a-5-percent-chance-of-avoiding-dangerous-global-warming-a-study-finds/?utm_term=.6d387390e4ca. [Accessed 29 August 2017].

Mooney, C. (2017b). *'New science suggests the ocean could rise more — and faster — than we thought'*, *Washington Post*, 26 October, [Online]. Available at: https://www.washingtonpost.com/news/energy-environment/wp/2017/10/26/new-science-suggests-the-ocean-could-rise-more-and-faster-than-we-thought/?utm_term=.729ae2882213. [Accessed 24 January 2018].

Moore, J. and Rees, W. E. (2013). 'Getting to One-Planet Living', in Worldwatch Institute (eds.), *State of the World 2013*, Island Press, Washington, DC, pp. 39–50.

Moorhead, J. (2015). *'How the Endcap Dictates What You Buy at the Grocery Store'*, The Kitchn, 11 August. [Online]. Available at: http://www.thekitchn.com/how-the-endcap-dictates-what-you-buy-222128. [Accessed 29 August 2017].

Mora, C., Dousset, B., Caldwell, I. R., *et al.* (2017). 'Global risk of deadly heat', *Nature Climate Change,* 7, pp. 501–506.

Morowitz, H. J. (1986). *Mayonnaise and the Origin of Life*, Berkley, New York.

Mufson, S. (2017a). *'ExxonMobil refineries are damaged in Hurricane Harvey, releasing hazardous pollutants'*, The Washington Post, 29 August. [Online]. Available at: https://www.washingtonpost.com/news/energy-environment/wp/2017/08/29/exxonmobil-refineries-damaged-in-hurricane-harvey-releasing-hazardous-pollutants/?utm_term=.a53d08cd4704. [Accessed 30 August 2017].

Mufson, S. (2017b). *'Texas chemical plant in critical condition, raising possibility of explosion',* The Washington Post, 29 August. [Online]. Available at: https://www.washingtonpost.com/news/wonk/wp/2017/08/29/texas-chemical-plant-in-critical-condition-raising-possibility-of-explosion/?utm_term=.df5c5e78ed44. [Accessed 30 August 2017].

Muller, A., Schader, C., Scialabba, N. E-H., *et al.* (2017). 'Strategies for feeding the world more sustainably with organic agriculture, *Nature Communications'*, 8, pp. 1290–1303.

Muller, T. (2008). *'Cuba: Emerging from the crisis'*, Oxfam America, 26 September. [Online]. Available at: https://www.oxfamamerica.org/explore/stories/cuba-emerging-from-the-crisis/. [Accessed 29 August 2017].

Murphy, D. J. and Hall, C. A. S. (2010). 'Year in review—EROI or energy return on (energy) invested', *Ann. N.Y. Acad. Sci.*, 1185, pp. 102–118.

Murphy, P. and Morgan, F. (2013). 'Cuba: Lessons from a Forced Decline', in Worldwatch Institute (eds.), *Is Sustainability Still Possible? State of the World 2013*, Island Press, Washington, DC, pp. 332–342.

Murphy, Jr., T. W. (2013). 'Beyond Fossil Fuels: Assessing Energy Alternatives', in Worldwatch Institute (eds.), *Is Sustainability Still Possible? State of the World 2013*, Island Press, Washington, DC, pp. 172–183.

Myhre, G., Shindell, D., Bréon, F. M., *et al.* (2013). 'Anthropogenic and Natural Radiative Forcing', in Stocker, T. F., Qin, D., Plattner, G. K., *et al.*, *Climate Change 2013: The Physical Science Basis. Contribution of Working Group I to the Fifth Assessment Report of the Intergovernmental Panel on Climate Change.* Cambridge University Press, Cambridge.

NAS. (2016). *Genetically Engineered Crops: Experiences and Prospects*, National Academies of Sciences, Engineering, and Medicine, The National Academies Press, Washington, DC.

NASA. (2017). *'The Relentless Rise of Carbon Dioxide'.* [Online]. Available at: https://climate.nasa.gov/climate_resources/24/. [Accessed 8 April 2017].

Nature. (2017). 'Extreme weather events are the new normal. Hurricane Harvey highlights the struggle to apply climate science', *Nature*, 548, p. 499.

Nelson, M. K. (2013). 'Protecting the Sanctity of Native Foods', in Worldwatch Institute (eds), *Is Sustainability Still Possible? State of the World 2013*, Island Press, Washington, DC, pp. 201–209.

Neslen, A. (2016). 'Glyphosate unlikely to pose risk to humans, UN/WHO study says', The Guardian, 16 May. [Online]. Available at: https://www.theguardian.com/environment/2016/may/16/glyphosate-unlikely-to-pose-risk-to-humans-unwho-study-says. [Accessed 29 August 2017].

New York Times. (2017). *'China and India Make Big Strides on Climate Change'*, Editorial, 22 May. [Online]. Available at: https://www.nytimes.com/2017/05/22/opinion/paris-agreement-climate-china-india.html. [Accessed 29 August 2017].

Next System Project. (2017). [Online]. Available at: https://thenextsystem.org/. [Accessed 23 August 2017].

Nierenberg, D. (2013). 'Agriculture: Growing Food—and Solutions', in Worldwatch Institute (eds.), *State of the World 2013*, Island Press, Washington, DC, pp. 190–200.

NRC. (2011). *'Warming World: impact by degree'*, [Online]. Available at: http://dels.nas.edu/resources/static-assets/materials-based-on-reports/booklets/warming_world_final.pdf. [Accessed 26 January 2018].

Oaklander, M. and Tweeten, L. (2016). *'Should You Be Eating Bugs Instead of Meat?'*, TIME, 14 April. [Online]. Available at: http://time.com/4292792/edible-bugs-protein/. [Accessed 26 January 2018].

O'Connor, J. (1988). 'Capitalism, Nature, Socialism: A Theoretical Introduction', *Capitalism Nature Socialism*, 1(1), p. 34.

O'Connor, M. (1991). 'Entropy, Structure, and Organisational Change', *Ecological Economics*, 3, pp. 95–122.

OECD. (2010). *'Energy Poverty: How to make modern energy access universal'*, Organization for Co-operation and Development, and International Energy Agency. [Online]. Available at: http://www.se4all.org/sites/default/files/l/2013/09/Special_Excerpt_of_WEO_2010.pdf. [Accessed 24 August 2017].

OECD. (2017). *'Family Database. Child poverty'*, Organization for Co-operation and Development. [Online] (Updated 5 January 2014). Available at: http://www.oecd.org/social/family/database. [Accessed 23 August 2017].

Oelkers, S. R., Gislason, E. S., Aradottir, B. S., *et al.* (2016). 'Rapid carbon mineralization for permanent disposal of anthropogenic carbon dioxide emissions', *Science*, 352(6291), pp. 1312–1314.

OilWatch. (2015). *'It is time to create Annex 0, Proposal for COP21, Paris, December 2015'*. [Online]. Available at: http://www.oilwatch.org/en/keep-the-oil-in-the-soil/679-it-is-time-to-create-annex-0. [Accessed 18 August 2017].

Oliver, R. (2008). *'All About: Food and Fossil Fuels'*, CNN, 17 March. [Online]. Available at: http://edition.cnn.com/2008/WORLD/asiapcf/03/16/eco.food.miles/. [Accessed 26 January 2018].

Opalka, B. (2009). *'U.S. Military Leads Green Charge'*, Renewable Energy World, 29 July. [Online]. Available at: http://www.renewableenergyworld.com/articles/2009/07/u-s-military-leads-green-charge.html. [Accessed 30 January 2018].

OSHA. (2018). *'Green Job Hazards: Solar Energy'*, [Online]. Available at: https://www.osha.gov/dep/greenjobs/solar.html. [Accessed 31 January 2018].

Oxfam. (2001). *'Cuba: Going Against Grain'*, Oxfam America, 1 June. [Online]. Available at: https://www.oxfamamerica.org/publications/cuba-going-against-the-grain/?searchterm=cuba%20going%20against%20the%20grain. [Accessed 26 January 2018].

Oxfam. (2008). *'Valley of the sugar mills'*, Oxfam America, 26 September. [Online]. Available at: https://www.oxfamamerica.org/explore/stories/valley-of-the-sugar-mills/. [Accessed 29 August 2017]. [Accessed 26 January 2018].

Parenti, C. (2017). *'If We Fail'*, Jacobin, 29 August. [Online]. Available at: https://jacobinmag.com/2017/08/if-we-fail. [Accessed 31 August 2017].

Parker, L. (2016). *'What You Need to Know About the World's Water Wars'*, National Geographic, 14 July. [Online]. Available at: http://news.nationalgeographic.com/2016/07/world-aquifers-water-wars/. [Accessed 29 August 2017].

Parkinson, G. (2015). *'Citigroup sees $100 trillion of stranded assets if Paris succeeds'*, RenewEconomy, 25 August. [Online]. Available at: http://reneweconomy.com.au/citigroup-sees-100-trillion-of-stranded-assets-if-paris-succeeds-13431/. [Accessed 25 August 2017].

Parkinson, G. (2016). *'The Myth About Renewable Energy Subsidies'*, CleanTechnica, 25 February. [Online]. Available at: https://cleantechnica.com/2016/02/25/the-myth-about-renewable-energy-subsidies/. [Accessed 29 August 2017].

Parks, J. (2014). 'Soil-Free Farming Grows Vegetables in the Desert', *Live Science,* January 24, [Online]. Available at: https://www.livescience.com/42835-soil-free-farming-grows-vegetables-in-the-desert.html. [Accessed 29 August 2017].

Parlapiano A. (2015). *'Climate Goals Pledged by China and the U.S.'*, New York Times. [Online]. (Updated 2 October 2015). Available at: https://www.nytimes.com/interactive/2014/11/12/world/asia/climate-goals-pledged-by-us-and-china-2.html. [Accessed 18 August 2017].

Parn, J, Verhoeven, J.T.A., Butterbach-Bahl, K., *et al.*, (2018). 'Nitrogen-rich organic soils under warm well-drained conditions are global nitrous oxide emission hotspots', *Nature Communications*, 9, 1135, DOI: 10.1038/s41467-018-03540-1.

Paton, C. (2012). *'Seawater Greenhouse: A new approach to restorative agriculture'*, Global Water Forum, 28 May. [Online]. Available at: http://www.globalwaterforum.org/2012/05/28/seawater-greenhouse-a-new-approach-to-restorative-agriculture/?pdf=4105. [Accessed 29 August 2017].

Payne, C. L. R., Scarborough, P., Rayner, M., *et al.* (2016). 'Are edible insects more or less 'healthy' than commonly consumed meats? A comparison using two nutrient profiling models developed to combat over- and undernutrition', *European Journal of Clinical Nutrition*, 70, pp. 285–291.

Pehl, M., Arvesen, A., Humpenöder, F., *et al.* (2017). 'Understanding future emissions from low-carbon power systems by integration of life-cycle assessment and integrated energy modelling', *Nature Energy*, 2, pp. 939–945.

Pellegrino, E., Bedini, S., Nuti, M., *et al.* (2018) 'Impact of genetically engineered maize on agronomic, environmental and toxicological traits: a meta-analysis of 21 years of field data', *Scientific Reports,* 8, 15 Feb, DOI: 10.1038/s41598-018-21284-2.

Penrose, R. (2011). *Cycles of Time,* Vintage, New York.

Perry, J. B. (2017). *'Jeffrey B. Perry'*. [Online]. Available at: http://www.jeffreyb-perry.net/. [Accessed 28 August 2017].

Perlo-Freeman, S., Fleurant, A., Wezeman, P. D., *et al.* (2015). *'Trends in World Military Expenditure'*, SIPRI Fact Sheet. [Online]. Available at: https://sipri.org/sites/default/files/EMBARGO%20FS1604%20Milex%202015.pdf. [Accessed 26 January 2018].

Peters, G. P., Andrew, R. M., Boden, T., *et al.* (2013). 'The Challenge to Keep Global Warming Below 2 °C', *Nature Climate Change*, 3, pp. 4–6.

Peterson, G., Allen, C. R., and Holling, C. S. (1998). 'Ecological Resilience, Biodiversity, and Scale, *Ecosystems'*, 1(1), January, pp. 6–18.

Pfeiffer, A., Millar, R., Hepburn, C., *et al.* (2016). 'The '2 ºC capital stock' for electricity generation: Committed cumulative carbon emissions from the electricity generation sector and the transition to a green economy', *Applied Energy,* 179, pp. 1395–1408.

Philpott, T. (2012). *'Time to Stop Worrying and Learn to Love Industrial Agriculture'*, Mother Jones, 2 May, [Online]. Available at: http://www.motherjones.com/food/2012/05/organic-vs-conventional-agriculture-nature/. [Accessed 29 August 2017].

Philpott, T. (2013). *'A Brief History of Our Deadly Addiction to Nitrogen Fertilizer'*, Mother Jones, 19 April. [Online]. Available at: http://www.motherjones.com/food/2013/04/history-nitrogen-fertilizer-ammonium-nitrate/. [Accessed 29 August 2017].

Philpott, T. (2016). 'The Gulf of Mexico Is About to Experience a 'Dead Zone' the Size of Connecticut', *Mother Jones*, June, [Online]. Available at: http://www.motherjones.com/environment/2016/06/gulf-mexico-braces-monsterous-dead-zone. [Accessed 29 August 2017].

Piketty, T. (2014). *Capital in the Twenty-First Century*, Harvard University Press, Cambridge.

Pimentel, D., Hepperly, P., Hanson, J., *et al.* (2005). 'Environmental, Energetic, and Economic Comparison of Organic and Conventional Farming Systems', *Bioscience,* 55(7), pp. 573–582.

Pollen, M. (2007). *Omnivore's Dilemma: A Natural History of Four Meals,* Penguin, New York.

Ponisio, L. C., M'Gonigle, L. K., Mace, K. C., *et al.* (2014). 'Diversification practices reduce organic to conventional yield gap', *Proc. of Royal Society B.,* 282: 20141396, DOI: 10.1098/rspb.2014.1396.

Poschman, H. (2005). *'Water Usage in the West'*, CSG (Council of State Governments). [Online]. Available at: http://www.csgwest.org/policy/WesternWaterUsage.aspx. [Accessed 24 August 2017].

Possner, A. and Caldeira, K. (2017). 'Geophysical potential for wind energy over the open oceans', *Proceedings of the National Academy of Sciences (U.S.A.),* 114(43), pp. 11338–11343.

Post Carbon Institute. (2017). [Online]. Available at: http://www.postcarbon.org/. [Accessed 23 August 2017].

Postel, S. (2013). 'Sustaining Freshwater and Its Dependents', in Worldwatch Institute (eds.), *State of the World 2013*, Island Press, Washington, DC, pp. 51–62.

PP FAQ. (2017). *'Precautionary Principle—FAQs'*, Science & Environmental Health Network. [Online]. Available at: http://www.sehn.org/ppfaqs.html. [Accessed 26 January 2018].

Prakash, C. S. (2000). *'Genetically Engineered Crops Can Feed the World!'*, 21st Century Science & Technology Magazine. [Online]. Available at: http://www.21stcenturysciencetech.com/articles/biotech.html. [Accessed 20 August 2017].

Prigogine, I. and Stengers, I. (1984). *Order out of Chaos,* Bantam Book, Toronto.

Princen, T., Manno, J. P. and Martin, P. (2013). 'Keep Them in the Ground: Ending the Fossil Fuel Era', in Worldwatch Institute (eds.), *State of the World 2013*, Island Press, Washington, DC, pp. 161–171.

Proops, J. L. R. (1987). 'Entropy, information and confusion in the social sciences', *Journal of Interdisciplinary Economics,* 1, pp. 225–242.

Purvis, G., Downey, L., Beever, D., *et al.* (2012). 'Development of a Sustainably-Competitive Agriculture, Agroecology and Strategies for Climate Change', *Sustainable Agriculture Reviews,* 8, pp. 35–65.

Qi, Y., Stern, N., Wu, T., *et al.* (2016). 'China's post-coal growth', *Nature Geoscience*, 9, pp. 564–566.

Quinn, M. (2006). *'The Power of Community: How Cuba Survived Peak Oil'*, Resilience (resilience.org), 25 February. [Online]. Available at: http://www.resilience.org/stories/2006-02-25/power-community-how-cuba-survived-peak-oil/. [Accessed 26 January 2018].

Raftery, A. E., Zimmer, A., Frierson, D. M. W., *et al.* (2017). 'Less than 2 °C warming by 2100 unlikely', *Nature Climate Change,* DOI: 10.1038/nclimateTE3352.

Raza, A., Friedel, J. K. and Bodner, G. (2012). 'Improving Water Use Efficiency for Sustainable Agriculture, Agroecology and Strategies for Climate Change', *Sustainable Agriculture Reviews,* 8, pp. 167–212.

Reck, B. K. and Graedel, T. E. (2012). 'Challenges in Metal Recycling', *Science,* 337, pp. 690–695.

Reganold, J. (2016). *'Can we feed 10 billion people on organic farming alone?'*, The Guardian, 14 August. [Online]. Available at: https://www.theguardian.com/sustainable-business/2016/aug/14/organic-farming-agriculture-world-hunger. [Accessed 26 January 2018].

Reganold, J. P. and Wachter, J. M. (2016). 'Organic Agriculture in the 21st Century', *Nature Plants,* 2, 15221, DOI: 10.1038/nplants.2015.221

Rein, S. (2016). 'It's the end of the world as we know it: militarism then and now', in Ehmsen, S. and Scharenberg, A. (eds.), *Rosa Remix*, Rosa Luxemburg Stiftung, New York, pp. 75–80.

Reisner, M. (1986). *Cadillac Desert,* Penguin Books, New York.

Reyes, O. (2015). *'Words without action: Seven takeaways from the Paris climate deal'*, 14 December. [Online]. Available at: http://www.redpepper.org.uk/words-without-action-seven-takeaways-from-the-paris-climate-deal/. [Accessed 26 January 2018].

Rifkin, J. (1981). *Entropy*, Viking Press, New York.

Rifkin, J. (1989). *Entropy* (Revised edition), Bantam Books, New York.

Rifkin, J. (2011). T*he Third Industrial Revolution*, Palgrave MacMillan, New York.

Rifkin, J. (2014). *The Zero Marginal Cost Society*, Palgrave MacMillan, New York.

Riley, T. (2017). *'Just 100 companies responsible for 71% of global emissions, study says'*, The Guardian, 10 July. [Online]. Available at: https://www.theguardian.com/sustainable-business/2017/jul/10/100-fossil-fuel-companies-investors-responsible-71-global-emissions-cdp-study-climate-change. [Accessed 29 August 2017]. [Accessed 26 January 2018].

Rivers, W. (2014). *'High and Wide: Income Inequality Gap in the District One of the Biggest in the U.S.'*, DC Fiscal Policy Institute, 20 March. [Online]. Available at: https://www.dcfpi.org/wp-content/uploads/2014/03/3-13-14-Income-Inequality-in-DC.pdf [Accessed 26 January 2018].

Robinson, K. S. (2013). 'Is it too late?', in Worldwatch Institute (eds.), *State of the World 2013*, Island Press, Washington, DC, pp. 374–380.

Robinson, W. I. (2004). *A theory of global capitalism*, Johns Hopkins Univ. Press, Baltimore.

Rockström, J., Gaffney, O., Rogelj, J., *et al.* (2017). 'A roadmap for rapid decarbonization', *Science,* 355(6331), pp. 1269–1271.

Rockström, J., Steffen, W., Noone, K., *et al.* (2009). 'A safe operating space for humanity', *Nature*, 461, pp. 472–475.

Rogelj, J., McCollum, D. L., O'Neill, B. C., *et al*, (2013). '2020 emissions levels required to limit warming to below 2 ºC', *Nature Climate Change,* 3, pp. 405–412.

Rohde, R. A. and Muller, R. A. (2015). 'Air Pollution in China: Mapping of Concentrations and Sources', *PLOS ONE,* 10(8), e0135749, DOI: 10.1371/journal. pone.0135749.

Rosen, J. W. (2017). *'The World Cuts Back on Coal, a Growing Appetite in Africa'*, National Geographic, 10 May. [Online]. Available at: http://news.nationalgeographic.com/2017/05/lamu-island-coal-plant-kenya-africa-climate/. [Accessed 26 January 2018].

Roser, M. and Ritchie, H. (2017a). *'Fertilizer and Pesticides'*, Our World In Data. [Online]. Available at: https://ourworldindata.org/fertilizer-and-pesticides/. [Accessed 10 August 2017].

Roser, M. and Ritchie, H. (2017b). *'Food per Person'*, Our World In Data. [Online]. Available at: https://ourworldindata.org/food-per-person/. [Accessed 20 August 2017].

Rosset, P. and Benjamin, M. (eds,) (1994). *The Greening of the Revolution: Cuba's Experiment with organic agriculture,* Ocean Press, North Melbourne, Australia.

Rothman, T. (1989). *Science a la Mode*, Princeton University Press, Princeton, New Jersey.

Rowe, J., Dempsey, J. and Gibbs, P. (2016). 'The Power of Fossil Fuel Divestment (and its Secret)', in Carroll, W. K. and Sarker, K. (eds.), *A World To Win: Contemporary Social Movements and Counter-Hegemony*, ARP Books, Winnipeg.

Ruddiman,W. F. (2003). 'The atmospheric greenhouse era began thousands of years ago', *Climate Change*, 61, pp. 261–293.

Russel, D. A. and Williams, G. G. (1977). 'History of Chemical Fertilizer Development', *Soil Sci. Soc. Am. Journal*, 41, pp. 260–265.

Sachs, J. D. (2005). 'Can Extreme Poverty be Eliminated?', *Scientific American*, September. [Online]. Available at: https://www.scientificamerican.com/article/can-extreme-poverty-be-el/. [Accessed 26 January 2018].

Sadler, R. and Connell, D. (2012). *'Global Distillation in an Era of Climate Change, Organic Pollutants Ten Years After the Stockholm Convention - Environmental and Analytical Update*', [Online]. Available at: https://www.intechopen.com/books/organic-pollutants-ten-years-after-the-stockholm-convention-environmental-and-analytical-update/global-distillation-in-an-era-of-climate-change. [Accessed 26 January 2018].

Sagoff, M. (1995). 'Carrying capacity and ecological economics', *BioScience,* 45, pp. 610–620.

Salleh, A. (ed). (2014). *Eco-Sufficiency & Global Justic*e, Pluto Press, London.

Salter, G. L. (2013). *'Thousands of Chemicals on the Market But No Rules to Test for Safety', Physicians for Social Responsibility, Environmental Health Policy Institute,* 19 November. [Online]. Available at: http://www.psr.org/environment-and-health/environmental-health-policy-institute/responses/thousands-of-chemicals-on-the-market.html?referrer=https://www.google.com/. [Accessed 10 May 2017].

Sandler, B. (1994). 'Grow or die: Marxist theories of capitalism and the environment', *Rethinking Marxism,* 7(2), pp. 38–57.

Sauvaget, C., Ramadas, K,, Fayette, J. M., *et al.* (2011). 'Socio-economic factors & longevity in a cohort of Kerala State, India', *Indian J Med Res*, 133(5), pp. 479–486.

Sawin, J. L., Seyboth, K. and Sverrisson, F. (2017). *'Renewables 2017 Global Status Report.'*, REN 21. [Online]. Available at: http://www.ren21.net/wp-content/uploads/2017/06/17-8399_GSR_2017_Full_Report_0621_Opt.pdf. [Accessed 26 August 2017].

Scheer, H. (2005). *The Solar Economy,* Earthscan, London.

Scheer, H. (2007). *Energy Autonomy. The economic, social and technological case for renewable energy,* Earthscan, London.

Scherr, S. J. and McNeely, J. A. (eds.). (2007). *Farming with Nature: The Science and Practice of Ecoagriculture*, Island Press, Washington, DC.

Schlömer, S., Bruckner, T., Fulton, L., *et al.* (2014). 'Annex III: Technology-specific cost and performance parameters', in Edenhofer, O., Pichs-Madruga, R., Sokona, Y., *et al.* (eds.), *Climate Change 2014: Mitigation of Climate Change. Contribution of Working Group III to the Fifth Assessment Report of the Intergovernmental Panel on Climate Change,* Cambridge University Press, New York.

Schlosser, E. (2001). *Fast Food Nation: The Dark Side of the All-American Meal*, Houghton Mifflin Company, Boston.

Schneider, C. and Banks, J. (2010). *'The Toll From Coal: An Updated Assessment of Death and Disease from America's Dirtiest Energy Source'*, Clean Air Task Force. [Online]. Available at: http://www.catf.us/resources/publications/files/The_Toll_from_Coal.pdf. [Accessed 26 January 2018].

Schramski, J. R., Gattie, D. K. and Brown, J. H. (2015). 'Human domination of the biosphere: Rapid discharge of the earth-space battery foretells the future of humankind', *Proc Natl Acad Sci (USA),* 112(31), pp. 9511–9517.

Schwartzman, D. (1992a). 'To Each According to Her Needs', *Dialogue & Initiative*, 4 , pp. 16–19.

Schwartzman, D. (1992b). 'A World Party, Vehicle of Global Green Left', *Ecosocialist Review,* Spring.

Schwartzman, D. (1996). 'Solar Communism', *Science & Society*, 60 (3), pp. 307–331.

Schwartzman, D. (2002). *Life, temperature, and the earth: The self-organizing biosphere*, Columbia University Press, New York.

Schwartzman, D. (2008). 'The Limits to Entropy: Continuing Misuse of Thermodynamics in Environmental and Marxist theory', *Science & Society,* 72(1), pp. 43–62.

Schwartzman, D. (2009a). 'Ecosocialism or Ecocatastrophe?', *Capitalism Nature Socialism,* 20(1), pp. 6–33.

Schwartzman, D. (2009b). 'Response to Næss and Høyer, *Capitalism', Nature Socialism,* 20(4), pp. 93–97.

Schwartzman, D. (2010). *'Review of "Who Will Build the Ark" by Mike Davis'.* [Online]. Available at: http://www.dcmetrosftp.org/newsletters/NL20100316.html#Ark. [Accessed 29 August 2017].

Schwartzman, D. (2011). 'Green New Deal: An Ecosocialist Perspective', *Capitalism Nature Socialism,* 22(3), pp. 49–56.

Schwartzman, D. (2012). 'A Critique of Degrowth and Its Politics', *Capitalism Nature Socialism,* 23(1), pp. 119–25.

Schwartzman, D. (2013). '4 Scenarios for 2050', *Capitalism Nature Socialism,* 23(1), pp. 49–53.

Schwartzman, D. (2014a). 'Revolutions that Made the Earth. Book Review', *Am. J. Phys.*, 82, pp. 529–530.

Schwartzman, D. (2014b). 'Ted Trainer and the simpler Way: a somewhat less sympathetic critique', *Capitalism Nature Socialism,* 25(2), pp. 112–117.

Schwartzman, D. (2014c). 'My Response to Trainer', *Capitalism Nature Socialism,* 25(4), pp. 109–115.

Schwartzman, D. (2014d). 'Is zero economic growth necessary to prevent climate catastrophe?', *Science & Society,* 78(2), pp. 235–240.

Schwartzman, D. (2015a). 'From the Gaia hypothesis to a theory of the evolving self-organizing biosphere', *Metascience,* 24(2), pp. 315–319.

Schwartzman, D. (2015b). *'COP21: Achievements and challenges to the climate justice movement'.* [Online]. Available at: http://www.cnsjournal.org/cop21-achievements-and-challenges-to-the-climate-justice-movement/. [Accessed 26 January 2018].

Schwartzman, D. (2015c). *'Restoring Ecosystems to Reverse Global Warming? A Critique of Biodiversity for a Livable Climate claims'*, 4 October. [Online]. Available at: http://solarutopia.org/wp-content/uploads/2015/12/Critique.pdf. [Accessed 29 August 2017].

Schwartzman, D. (2016a). 'Beyond eco-catastrophism: the conditions for solar communism', in Panitch, L. and Albo, G. (eds.), *Socialist Register 2017*, Monthly Review Press, New York, pp. 143–160.

Schwartzman, D. (2016b). *'Should we reject Negative Emissions Technologies except for organic agriculture?'*, CNS Journal, 5 February. [Online]. Available at: http://www.cnsjournal.org/should-we-reject-negative-emissions-technologies-except-for-organic-agriculture/. [Accessed 29 August 2017].

Schwartzman, D. (2016c). 'How Much and What Kind of Energy Does Humanity Need?', *Socialism and Democracy*, 30(2), pp. 97–120.

Schwartzman, D. (2017). *'Report from the First Ecosocialist International Launch Meeting in Venezuela'*, Our Place in the World: A Journal of Ecosocialism, 23 December. [Online]. Available at: https://forhumanliberation.blogspot.com/2017/12/2783-report-from-first-ecosocialist.html. [Accessed 27 January 2018].

Schwartzman, D. and Saul, Q. (2015). 'An Ecosocialist Horizon for Venezuela: A Solar Communist Horizon for the World', *Capitalism Nature Socialism,* 26(3), pp. 14–30.

Schwartzman, D. and Schwartzman P. (2013). 'A rapid solar transition is not only possible, it is imperative!', *African Journal of Science, Technology, Innovation and Development,* 5(4), pp. 297–302.

Schwartzman, D. and Schwartzman P. (2017). Review of Robert Biel's Sustainable Food Systems, *Journal of Labour and Society,* 20(2), pp. 272–276.

Schwartzman, P. and Schwartzman, D. (2011). *'A Solar Transition is Possible'*, Institute for Policy Research and Development, [Online]. Available at: http://solarutopia.org/wp-content/uploads/2013/04/A-Solar-Transition-is-Possible_new.pdf. [Accessed 26 January 2018].

Schwartzman, P. and Schwartzman. D. (2017). Solar Utopia website. [Online]. Available at: http://solarutopia.org/. [Accessed 26 January 2018].

Schwartzman, P., Schwartzman, D., and Zhang, X. (2016). *'Climatic implications of a rapid wind/solar transition'*, Arxiv.org, 22 March, [Online]. Available at: https://arxiv.org/pdf/1603.06929.pdf. [Accessed 26 January 2018].

Schwenke, D. C. (2016). Increasing dietary fruits and vegetables to reduce healthcare costs, *Current Opinion in Lipidology,* 27(4), pp. 422–423.

Science News. (2015). *'New cathode material creates possibilities for sodium-ion batteries'*, Science Daily, 23 September. [Online]. Available at: http://www.sciencedaily.com/releases/2015/09/150923182801.htm. [Accessed 29 August 2017].

SEFA. (2017). *Sustainable Energy For All.* [Online]. Available at: http://se4all.org. [Accessed 10 August 2017].

SEJUP. (1994). *'Sterilization of Women in Brazil'*, Servico Brasileiro de Justica e Paz, 17 November. [Online]. Available at: http://pangaea.org/street_children/latin/brazil1.htm. [Accessed 29 August 2017].

Sen, A. (1993). The Economics of Life and Death, *Scientific American*, May, pp. 40–47.

Sgouridis, S., Csala, D. and Ugo Bardi, U. (2016). 'The sower's way: quantifying the narrowing net-energy pathways to a global energy transition', *Environmental Research Letters*, 11, 094009, DOI: 10.1088/1748-9326/11/9/094009.

Shahan, Z. (2012). *'Oil & Gas – Over 13 Times More in Historical Subsidies than Clean Energy'*, Clean Technica, 3 August. [Online]. Available at: https://cleantechnica.com/2012/08/03/oil-gas-over-13-times-more-in-historical-subsidies-than-clean-energy/. [Accessed 29 August 2017].

Shahan, Z. (2013). *'Over 3 times More Green Jobs per $1 Invested Than Fossil Fuel or Nuclear Jobs', Clean Technica,* 20 March. [Online]. Available at: https://cleantechnica.com/2013/03/20/over-3-times-more-green-jobs-per-million-than-fossil-fuel-or-nuclear-jobs/. [Accessed 23 August 2017].

Sharfiee, S. and Topal, E. (2009). When will fossil fuel reserves be diminished?, *Energy Policy*, 37(1), January, pp. 181–189.

SHEN. (2017). *'Precautionary Principle FAQs',* Science and Environmental Health Network. [Online]. Available at: http://www.sehn.org/ppfaqs.html. [Accessed 24 August 2017].

Shiva, V. (2008). *Soil Not Oil: Environmental Justice in an Age of Climate Crisis*, South End Press, Boston.

Shockman, E. (2016). *'What China's successful reforestation program means for the rest of the world'*, Public Radio International, 27 June. [Online]. Available at: https://www.pri.org/stories/2016-06-27/what-chinas-successful-reforestation-program-means-rest-world. [Accessed 29 August 2017].

Shukla, P. (2017) 'The 12 Most Pesticide-Contaminated Fruits and Vegetables of 2015'. 21 June. [Online]. Available at: https://food.ndtv.com/food-drinks/the-12-most-pesticide-contaminated-fruits-and-vegetables-of-2015-752555. [Accessed 8 April 2018].

Shuman, M. (2015). *The Local Economy Solution: How Innovative, Self-Financing "Pollinator" Enterprises Can Grow Jobs and Prosperity*, Chelsea Green, White River Junction, Vermont.

Singh, M. S., Kuang, Z., Maloney, E. D., et al. (2017). 'Increasing potential for intense tropical and subtropical thunderstorms under global warming',

Proceedings of the National Academy of Sciences (U.S.A.), 114(44), pp. 11657–11662, DOI: 10.1073/pnas.1707603114.

SIPRI. (2015). *'Fact Sheet'*, April. [Online]. Available at: https://www.sipri.org/. [Accessed 23 August 2017].

Skirbekk, V. (2008). Fertility trends by social status, *Demographic Research*, 18, pp. 145–180.

Smil, V. (1992). *General Energetics*, Wiley, New York.

Smil, V. (2003). *Energy at the Crossroads,* MIT Press, Cambridge.

Smil, V. (2008). *Energy in nature and society*, MIT Press, Cambridge.

Smith, G. (2012). *Nuclear Roulette*, Chelsea Green Publishing, White River Junction, Vermont.

Smith, P., Davis, S. J., Creutzig, F., *et al.* (2016). 'Biophysical and economic limits to negative CO_2 emissions', *Nature Climate Change,* 6, pp. 42–50.

Smith, R. (2011). 'Green capitalism: the god that failed', *Real-World Economics Review,* 56, pp. 112–144.

Smith, R. (2016). *Green Capitalism. The God that Failed*, World Economics Association and College Publications, Bristol.

Smolin, L. (1997). *The Life of the Cosmos*, Oxford, New York.

Smolin, L. (2013). *Time Reborn*, Houghton Mifflin Harcourt, Boston.

Sole, R. V. and Bascompte, J. (2006). *Self-organization in complex ecosystems*, Princeton University Press, Princeton.

Solon, O. (2015) Bees put to work lugging pesticides to flowers, *New Scientific*, 228(3045), p. 13.

Sood, J. (2013). *'Synthetic pesticide use is on the decline; bio-pesticide use is up'*, Down to Earth, 9 August. [Online]. Available at: http://www.downtoearth.org.in/news/synthetic-pesticide-use-is-on-the-decline-biopesticide-use-is-up-41891. [Accessed 29 August 2017].

Sorrell, S. (2015). Reducing energy demand: A review of issues, challenges and approaches. *Renewable and Sustainable Energy Reviews,* 47, pp. 74–82.

Sotos-Prieto, M., Bhupathiraju, S., Mattei, J., *et al.* (2017). 'Association of Changes in Diet Quality with Total and Cause-Specific Mortality', *New England Journal Medicine*, 377, pp. 143–53.

Spector, D. (2014). *'Scientists May Have Finally Pinpointed What's Killing All The Honeybees'*, Business Insider, 13 May. [Online]. Available at: http://www.businessinsider.com/harvard-study-links-pesticides-to-colony-collapse-disorder-2014-5. [Accessed 29 August 2017].

Spector, J. (2017). *'A Plan to Turn Shuttered Coal Plants Into Cutting-Edge Compressed Air Storage'*, 12 April. [Online]. Available at: https://www.greentechmedia.com/articles/read/plan-to-turn-shuttered-coal-plants-cutting-edge-compressed-air-storage. [Accessed 29 August 2017].

Steffen, W., Broadgate, W., Deutsch, L., et al. (2015). 'The trajectory of the Anthropocene: The Great Acceleration', *The Anthropocene Review*, DOI: 10.1177/2053019614564785.

Stein, J. (2012). *'Give U.S. a mandate for what America needs: a Green New Deal'*, The Guardian, 14 October. [Online]. Available at: https://www.theguardian.com/commentisfree/2012/oct/14/mandate-america-green-new-deal. [Accessed 23 August 2017].

Steingraber, S. (2003). *Having Faith,* The Berkley Publishing Group, New York.

Steingraber, S. (2010). *Living Downstream: An Ecologist's Personal Investigation of Cancer and the Environment,* Da Capo Press, Boston.

Steph. (2014). *'Detroit farm flowers for your Detroit wedding',* Love in the D, 12 September. [Online]. Available at: http://www.loveinthed.com/2014/09/12/detroit-farm-flowers-for-your-detroit-wedding/. [Accessed 24 August 2017].

Strauss, E. (2016). *'Turns Out the Infant Mortality Rate in the U.S. is Not That Bad—If You are Rich'*, Slate, 7 June. [Online]. Available at: http://www.slate.com/blogs/xx_factor/2016/06/07/the_u_s_infant_mortality_rate_is_high_due_to_wealth_disparities_according.html. [Accessed 29 August 2017].

Swanson, N. L., Leu, A., Abrahamson, J., *et al.* (2014) *'Genetically engineered crops, glyphosate and the deterioration of health in the United States of America,'* Journal of Organic Systems, 9(2), pp. 6–37. [Online]. Available at: http://www.organic-systems.org/journal/92/JOS_Volume-9_Number-2_Nov_2014-Swanson-et-al.pdf. [Accessed 25 February 2018].

Sweeney, S. (2015). Green Capitalism Won't Work, *New Labor Forum*, 24(2), pp. 12–17.

Sweeney, S. and Treat, J. (2017). 'Energy Transition: Are We Winning?', Working Paper No. 9, Trade Unions for Energy Democracy, Rosa Luxemburg Stiftung, Murphy Institute at CUNY, New York.

Sweeney, S., Benton-Connell, K. and Skinner, L. (2015). 'Power to the People', Working Paper No. 4, Trade Unions for Energy Democracy, Rosa Luxemburg Stiftung, Murphy Institute at CUNY, New York.

Taber, S. (2016). *'7 Facts That Will Make You Rethink the 'Sterility' of Hydroponics'*, Bright Agrotech, 13 May. [Online]. Available at: http://blog.

brightagrotech.com/7-facts-that-will-make-you-rethink-the-sterility-of-hydroponics. [Accessed 29 August 2017].

Tanuro, D. (2008). *'Fundamental Inadequacies of Carbon Trading for the Struggle Against Climate Change'*, Climate and Capitalism, 23 March. [Online]. Available at: http://climateandcapitalism.com/2008/03/23/carbon-trading-an-ecosocialist-critique/. [Accessed 29 August 2017].

Tenenbaum, D. J. (2008). Food vs. Fuel: Diversion of Crops Could Cause More Hunger, *Environ Health Perspect*, 116(6), pp. A254–A257.

Thompson, D. (2011). *'How We Spend Money—in China, India, Russia, Egypt, Brazil and the US'*, The Atlantic, 25 March. [Online]. Available at: https://www.theatlantic.com/business/archive/2011/03/how-we-spend-money-in-china-india-russia-egypt-brazil-and-the-us/73001/. [Accessed 29 August 2017].

Thompson, I., Mackey, B., McNulty, S., *et al.* (2009). *'Forest Resilience, Biodiversity, and Climate Change: A synthesis of the biodiversity/resilience/stability relationship in forest ecosystems'*, Secretariat of the Convention on Biological Diversity, Technical Series no. 43. [Online]. Available at: https://pdfs.semanticscholar.org/4b01/a1a24ef80d4a6e71e7267ca91cc64478aed2.pdf. [Accessed 30 August 2017].

Thompson, J., Hodgkin, T., Kuresi, A. K., *et al.* (2007). 'Biodiversity in Agroecoystems', in Scherr, S. J. and McNeely, J. (eds.), *Farming with Nature: The Science and Practice of Ecoagriculture*, Island Press, Washington D.C.

Tienhaara, K. (2014). 'Varieties of green capitalism: economy and environment in the wake of the global financial crisis', *Environmental Politics,* 23(2), DOI: 10.1080/09644016.2013.821828.

Tilman, D., Cassman, K. G., Matson, P. A., *et al.* (2002). 'Agricultural sustainability and intensive production practices', *Nature,* 418, pp. 671–677.

Tokar, B. (2014). *Toward Climate Justice: Perspectives on Climate Crisis and Social Change,* Norway: New Compass Press.

Tollefson, J. and Weiss, K. R. (2015). 'Nations adopt historic global climate accord', *Nature*, 528, pp. 315–316.

Touliatos, D., Dodd, I. C. and McAinsh, M. (2016). 'Vertical farming increases lettuce yield per unit area compared to conventional horizontal hydroponics', *Food and Energy Security*, 5(3), pp. 184–191.

Trade Unions for Energy Democracy. (2017). [Online]. Available at: http://unionsforenergydemocracy.org/. [Accessed 29 August 2017].

Trainer, T. (2011). 'The radical implications of a zero growth economy', *Real-World Economics Review,* 57, pp. 71–82.

Trainer, T. (2014). 'Reply to David Schwartzman on the Simpler Way and Renewable Energy', *Capitalism Nature Society,* 25(4), pp. 102–108.

Trautmann, N. M., Porter, K. S. and Wagenet, R. J. (2012). *'Modern Agriculture: Its Effects on the Environment'*, [Online]. Available at: http://psep.cce.cornell.edu/facts-slides-self/facts/mod-ag-grw85.aspx. [Accessed 29 August 2017].

Treves, A., Artelle, K. A., Darimont, C. T., *et al.* (2018). 'Intergenerational equity can help to prevent climate change and extinction, *Nature Ecology & Evolution*, DOI: 10.1038/s41559-018-0465-y.

Trisos C. H., Amatulli, G., Gurevitch, J. *et. al.* (2018). Potentially dangerous consequences for biodiversity of solar geoengineering implementation and termination', *Nature Ecology & Evolution*, DOI: 10.1038/s41559-017-0431-0.

Turse, N. (2009). *The Complex: How the Military Invades Our Everyday Lives,* Metropolitan Books, New York.

TWN & SOCLA (2015). *'Agroecology: Key Concepts, Principles & Practices'*, Third World Network (TWN) and Sociedad Cientifica Latinoamericana de Agroecologia (SOCLA). [Online]. Available at: https://www.twn.my/title2/books/pdf/Agroecologycomplete1.pdf. [Accessed 26 January 2018].

Ulmer, A. and Parraga, M. (2014). *'Venezuela's crude imports show PDVSA picks pragmatism over politics',* Reuters, 27 October. [Online]. Available at: http://www.reuters.com/article/us-venezuela-oil-imports-analysis-idUSKBN0IG17F20141027. [Accessed 26 January 2018].

UNCTAD. (2013). *'Wake Up Before It Is Too Late: Make Agriculture Truly Sustainable Now For Food Security in a Changing Climate'*, United Nations Conference of Trade and Development. [Online]. Available at: http://unctad.org/en/PublicationsLibrary/ditcted2012d3_en.pdf. [Accessed 26 January 2018].

UNEP. (2012). *'Thawing of permafrost expected to cause significant additional global warming, not yet accounted for in climate predictions'*, United Nations Environmental Programme, 27 November. [Online]. Available at: http://web.unep.org/newscentre/thawing-permafrost-expected-cause-significant-additional-global-warming-not-yet-accounted-climate. [Accessed 26 January 2018].

UNEP. (2014). *'Year Book 2014 emerging issues update. Air Pollution: World's Worst Environmental Health Risk',* [Online]. Available at: http://staging.unep.org/yearbook/2014/PDF/UNEP_YearBook_2014.pdf. [Accessed 26 January 2018].

UN Framework Convention on Climate Change. (2015). *'Adoption of the Paris agreement'*, 12 December. [Online]. Available at: https://unfccc.int/resource/docs/2015/ cop21/eng/l09.pdf. [Accessed 28 August 2017].

UNHCR. (2016). *'Frequently Asked Questions on Climate Change and Disaster Displacement'*, The UN Refugee Agency, November. [Online]. Available at: http://www.unhcr.org/en-us/protection/environment/581870fc7/frequently-asked-questions-climate-change-disaster-displacement.html. [Accessed 26 January 2018].

UNICEF. (2013). *'Overall funding trends'*, United Nations Children's Fund. [Online]. Available at: http://www.unicef.org/appeals/funding_trends.html. [Accessed 23 August 2017].

United Nations. (2015). *'World Population Prospects The 2015 Revision'*, Department of Economic and Social affairs. [Online]. Available at: https://esa.un.org/unpd/wpp/publications/files/key_findings_wpp_2015.pdf. [Accessed 20 May 2017].

United Nations. (2017). *'World Population Prospects: 2017 Revision'*, Department of Economic and Social Affairs. [Online]. Available at: https://esa.un.org/unpd/wpp/. [Accessed 24 August 2017].

Urbina, I. (2013). *'Think Those Chemicals Have Been Tested?'*, New York Times, 13 April. [Online]. Available at: http://www.nytimes.com/2013/04/14/sunday-review/think-those-chemicals-have-been-tested.html?mcubz=3. [Accessed 10 May 2017].

USDA. (2014a). *'U.S. Farms and Farmers: Preliminary Report Highlights'*, United States Department of Agriculture, February. [Online]. Available at: https://www.agcensus.usda.gov/Publications/2012/Preliminary_Report/Highlights.pdf. [Accessed 26 January 2018].

USDA. (2014b). *'Farms & Farmland'*, United States Department of Agriculture, September. [Online]. Available at: https://www.agcensus.usda.gov/Publications/2012/Online_Resources/Highlights/Farms_and_Farmland/Highlights_Farms_and_Farmland.pdf. [Accessed 26 January 2018].

USDA. (2017a). *'Fertilizer Use & Markets'*, United States Department of Agriculture, Economic Research Service. [Online]. Available at: http://www.ers.usda.gov/topics/farm-practices-management/chemical-inputs/fertilizer-use-markets.aspx. [Accessed 26 January 2018].

USDA. (2017b). *'Food Prices and Spending'*. [Online]. Available at: https://www.ers.usda.gov/data-products/ag-and-food-statistics-charting-the-essentials/food-prices-and-spending/. [Accessed 26 January 2018].

USDA. (2017c). *'Organic Market Overview'*, United States Department of Agriculture, Economic Research Service. [Online]. Available at: https://

www.ers.usda.gov/topics/natural-resources-environment/organic-agriculture/organic-market-overview/. [Accessed 29 August 2017].

USDA. (2017d). *'Corn and Other Feed Grains'*, United States Department of Agriculture. [Online]. Available at: https://www.ers.usda.gov/topics/crops/corn/background.aspx. [Accessed 29 August 2017].

USA DoD. (2015). *'DoD Releases Report on Security Implications of Climate Change'* Department of Defence, 29 July. [Online]. Available at: https://www.defense.gov/News/Article/Article/612710. [Accessed 28 August 2017].

USA DoE. (2014). *'How Much Do You Spend'*, Department of Energy, 2 July. [Online]. Available at: https://energy.gov/articles/how-much-do-you-spend. [Accessed 28 August 2017].

US EIA. (2015). *'Average Operating Heat Rate for Selected Energy Sources'*, Energy Information Administration. [Online]. Available at: https://www.eia.gov/electricity/annual/html/epa_08_01.html. [Accessed 28 August 2017].

US EIA. (2017). *'Annual Energy Outlook 2017'*, Energy Information Administration, 5 January. [Online]. Available at: https://www.eia.gov/outlooks/aeo/pdf/0383(2017).pdf. [Accessed 28 August 2017].

US EIA. (2018). *'How much carbon dioxide is produced when different fuels are burned?'* Energy Information Administration, [Online]. Available at: https://www.eia.gov/tools/faqs/faq.php?id=73&t=11. [Accessed 28 January 2018].

US EPA. (2017). *'Pesticide Registration'*, Environment Protection Agency. [Online]. Available at: https://www.epa.gov/pesticide-registration/inert-ingredients-regulation. [Accessed 27 August 2017].

USGS. (2010). *'Total Water Use in the United States, 2010'*, United States Geological Survey. [Online]. Available at: https://water.usgs.gov/edu/wateruse-total.html. [Accessed 14 August 2017].

USGS. (2017). *'The Water Content of Things'*, United States Geological Survey. [Online]. Available at: https://water.usgs.gov/edu/activity-watercontent.php. [Accessed 24 August 2017].

Vandermeer, J. and Perfecto, I. (2005) *Breakfast of Biodiversity: The Truth About Rainforest Destruction (second edition)*, Food First, Oakland, California.

Van Huis, A., Van Itterbeeck, J., Klunder, H., *et al.* (2013). *'Edible insects: future prospects for food and feed security'*, FAO Forestry Paper 171, [Online]. Available at: http://www.fao.org/docrep/018/i3253e/i3253e.pdf. [Accessed 28 January 2018].

Van Leeuwen, J. W. S. (2013). *'Nuclear Power Insights'*. [Online]. Available at: http://www.stormsmith.nl/insight-items.html. [Accessed 29 August 2017].

Van Noppen, T. (2013). *'Dirty Water: Can U.S. Clean Up Its Act?'*, Live Science, 11 April. [Online]. Available at: https://www.livescience.com/28669-dirty-water-report.html. [Accessed 29 August 2017].

Van Vuuren, D. P., Hof, A. F., van Sluisveld, M. A. E., *et al.* (2017). Open discussion of negative emissions is urgently needed, *Nature Energy,* 2, pp. 902–904.

Volk, T. (1998). *Gaia's Body: Toward a Physiology of Earth,* Springer-Verlag, New York.

von Weiszacher, E. U., Hargroves, C., Smith, M. H., *et al.* (2009). *Factor Five. Transforming the Global Economy Through 80% Improvements in Resource Productivity,* Earthscan, London.

Wagar, W. W. (1989). *A Short History of the Future (Revised Editions 1992, 1999),* University of Chicago Press, Chicago.

Wagar, W. W. (2001). *Memoirs of the Future*, Global Academic Publishing.

Wall, D. (2005). *Babylon and Beyond*, Pluto Press, London.

Wallis, V. (2015). 'Intersectionality's Binding Agent', *New Political Science,* 37(4), pp. 604–619.

Walsh, B., Ciais, P., Janssens, I. A., *et al.* (2017). 'Pathways for balancing CO_2 emissions and sinks', *Nature Comm.*, 8, 14856, DOI: 10.1038/ncomms14856.

Warrick, J. (2016). *'Wind, solar power soaring in spite of bargain prices for fossil fuels'*, The Washington Post, 30 December. [Online]. Available at: https://www.washingtonpost.com/national/health-science/wind-solar-power-soar-in-spite-of-bargain-prices-for-fossil-fuels/2015/12/30/754758b8-af19-11e5-9ab0-884d1cc4b33e_story.html?utm_term=.f3a61dcc3a57. [Accessed 26 January 2018].

Washington State University. (2008). USDA, [Online]. Available at: http://wsm.wsu.edu/researcher/WSMaug11_billions.pdf. [Accessed 26 January 2018].

Weber, G. W., Woodhouse, M., Kurtz, S., *et al.* (2017). Terawatt-scale photovoltaics: Trajectories and challenges, *Science,* 356, pp. 141–143.

Weinrub, A. (2010). *Community Public Power. Decentralized Renewable Energy in California,* Sierra Club California Energy-Climate Committee, Sacramento; Local Clean Energy Alliance, Oakland.

WEC. (2016a). *'World Energy Resources: Solar'*, World Energy Council. [Online]. Available at: https://www.worldenergy.org/wp-content/uploads/2017/03/WEResources_Solar_2016.pdf. [Accessed 26 January 2018].

WEC. (2016b). '*World Energy Resources 2016*', World Energy Council. [Online]. Available at: https://www.worldenergy.org/publications/2016/world-energy-resources-2016/. [Accessed 23 August 2017].

Wetter, K. J. and Zundel, T. (eds.) (2017). 'Big Bad Fix: The Case against Climate Geoengineering', Heinrich Boll Stiftung, [Online]. Available at: https://www.boell.de/en/2017/12/01/big-bad-fix-case-against-geoengineering, [Accessed 25 January 2018].

Wezel, A. (ed). (2017). *Agroecological Practices for Sustainable Agriculture: Principles, Applications, and Making the Transition*, World Scientific, Singapore.

Wezel, A. and David, C. (2012). 'Agroecology and the Food System', *Agroecology and Strategies for Climate Change, Sustainable Agriculture Reviews,* 8, pp. 17–34.

Wezel, A., Bellon, S., Dore, T., *et al.* (2009). 'Agroecology as a science, a movement and a practice: A review', *Agronomy Sustainable Development,* DOI: 10.1051/agro/2009004.

WFP. (2017). *'Cuba'*, World Food Program. [Online]. Available at: http://www1.wfp.org/countries/cuba. [Accessed 29 August 2017].

WHO. (2004). *'The Impact of Pesticides on Health: Preventing Intentional and Unintentional Deaths from Pesticide Poisoning', World Health Organization.* [Online]. Available at: http://www.who.int/mental_health/prevention/suicide/en/PesticidesHealth2.pdf. [Accessed 29 August 2017].

WHO. (2014). *'7 million premature deaths annually linked to air pollution'*, World Health Organization, 25 March. [Online]. Available at: http://www.who.int/mediacentre/news/releases/2014/air-pollution/en/. [Accessed 23 August 2017].

WHO. (2016). *'Life Expectancy: Data by Country'*, World Health Organization. [Online] (Updated 6 June 2016). Available from: http://apps.who.int/gho/data/node.main.688. [Accessed 23 August 2017].

WHO. (2017a). *'Chronic Disease and Health Promotion'*, World Health Organisation. [Online]. Available from: http://www.who.int/chp/chronic_disease_report/part1/en/index11.html. [Accessed 23 August 2017].

WHO. (2017b). *'Density of Physicians'*, World Health Organization. [Online]. Available from: http://gamapserver.who.int/gho/interactive_charts/health_workforce/PhysiciansDensity_Total/atlas.html. [Accessed 23 August 2017].

Wikipedia. (n.d.a). *'Congress of Industrial Organizations'*, [Online]. Available at: http://en.wikipedia.org/wiki/Congress_of_Industrial_Organizations. [Accessed 23 August 2017].

Wikipedia. (n.d.b). *'New Deal'*. [Online]. Available at http://en.wikipedia.org/wiki/New_Deal. [Accessed 23 August 2017].

Wikipedia. (n.d.c.). *'From each according to his ability, to each according to his needs'*. [Online]. Available at: https://en.wikipedia.org/wiki/From_each_according_to_his_ability,_to_each_according_to_his_needs. [Accessed 27 August 2017).

Wikipedia. (2016a). *'List of countries by life expectancy'*. [Online]. Available at: https://en.wikipedia.org/wiki/List_of_countries_by_life_expectancy. [Accessed 27 August 2017].

Wikipedia. (2016b). *'List of countries by income equality'*. [Online]. Available at: http://en.wikipedia.org/wiki/List_of_countries_by_ income_equality. [Accessed 27 August 2017].

Wikipedia. (2017a). *'Feed-in tariffs in Germany'*. [Online]. Available at: https://en.wikipedia.org/wiki/Feed-in_tariffs_in_Germany. [Accessed 26 August 2017].

Wikipedia. (2017b). *'Nuclear power in China'*. [Online]. Available at: https://en.wikipedia.org/wiki/Nuclear_power_in_China. [Accessed 26 August 2017].

Wikipedia. (2017c). *'Oil Reserves in Venezuela'*, Wikipedia. [Online]. Available at: https://en.wikipedia.org/wiki/Oil_reserves_in_Venezuela. [Accessed 29 August 2017].

Wilkinson, R. and Pickett, K. (2009). *The Spirit Level Why Equality is Better for Everyone*, Penguin Books, London.

Williams, C. (2010). *Ecology and socialism,* Haymarket Books, Chicago.

Wilson, E. O. (1992). The Diversity of Life. Belknap Press, Cambridge, MA.

Wong, E. (2016). *'Coal Burning Causes the Most Air Pollution Deaths in China, Study Finds'*, New York Times, 17 August. [Online]. Available at: https://www.nytimes.com/2016/08/18/world/asia/china-coal-health-smog-pollution.html?_r=0. [Accessed 29 August 2017].

Woods, J., Williams, A., Hughes, J. K., *et al.* (2010). 'Energy and the food system', *Philosophical Transactions of the Royal Society, B,* 365, pp. 2991–3006.

Wootton, D. (2015). *The Invention of Science*, HarperCollins, New York.

World Bank. (2013). *'GINI index (World Bank estimate)—Country Ranking'*. [Online]. Available at: https://www.indexmundi.com/facts/indicators/SI.POV.GINI/rankings. [Accessed 29 August 2017].

World Bank. (2014). *'Energy use (kg of oil equivalent per capita'*. [Online]. Available at: https://data.worldbank.org/indicator/EG.USE.PCAP.KG.OE. [Accessed 30 August 2017].

World Bank. (2015). *'Energy imports, net (% of energy use)'*. [Online]. Available at: http://data.worldbank.org/indicator/EG.IMP.CONS.ZS. [Accessed 29 August 2017].

World Bank. (2015b). *'Population Statistics for Nations'*. [Online]. Available at: http://statisticstimes.com/population/countries-by-population.php. [Accessed 29 August 2017].

World Bank. (2015c). *'GDP per Capita (current US$)'*. [Online]. Available at: http://data.worldbank.org/indicator/NY.GDP.PCAP.CD. [Accessed 29 August 2017].

World Bank. (2016a). *'Urban population (% of total)'*. [Online]. Available at: http://data.worldbank.org/indicator/SP.URB.TOTL.IN.ZS. [Accessed 29 August 2017].

World Bank. (2016b). *'Population growth (annual %)'*. [Online]. Available at: http://data.worldbank.org/indicator/SP.POP.GROW. [Accessed 20 August 2017].

World Bank. (2017). *'Mortality rate, infant (per 1,000 live births)'*. [Online]. Available at: http://data.worldbank.org/indicator/SP.DYN.IMRT.IN?year_high_desc=false. [Accessed 29 August 2017].

World Energy Outlook. (2012). *'Executive Summary'*. [Online]. Available at: http://www.worldenergyoutlook.org. [Accessed 29 August 2017].

World Population History. (2018). [Online]. Available at: http://www.worldpopulationhistory.org. [Accessed 8 April 2018].

United Nations. (2017). 'World Population Prospects: The 2017 Revision', Department of Economic and Social Affairs. June 21 2017. [Online]. Available at: https://www.un.org/development/desa/publications/world-population-prospects-the-2017-revision.html. [Accessed 26 January 2018].

WPR. (2017). *'Total Population By Country 2017'*, World Population Review. [Online]. Available at: http://worldpopulationreview.com/countries/. [Accessed 29 August 2017].

Wright, C. and Nyberg, D. (2018). *'We can't rely on corporations to save us from climate change'*, 30 January, London School of Economic Business Review. [Online]. Available at: http://blogs.lse.ac.uk/businessreview/2018/01/30/we-cant-rely-on-corporations-to-save-us-from-climate-change/. [Accessed 31 January 2018].

Wright, E. O. (2015). *'Why Class Matters'*, Jacobin, 23 December. [Online]. Available at: https://www.jacobinmag.com/2015/12/socialism-marxism-democracy-inequality-erik-olin-wright. [Accessed 29 August 2017].

Wright, E. O. (2016). *'How to Think About (And Win) Socialism'*, Jacobin, 27 April. [Online]. Available at: https://www.jacobinmag.com/2016/04/erik-olin-wright-real-utopias-capitalism-socialism. [Accessed 29 August 2017].

Wu, M. and Yiwen, C. (2011). *'Consumptive Water Use in the Production of Ethanol and Petroleum Gasoline (Updated July 2011)'*. [Online]. Available at: https://greet.es.anl.gov/files/consumptive-water. [Accessed 29 August 2017].

Xua, Y. and Ramanathan, V. (2017). 'Well below 2 °C: Mitigation strategies for avoiding dangerous to catastrophic climate changes', *Proceedings of the National Academy of Sciences (USA)*, 114(39), pp. 10315–10323.

Yang, X. J., Hu, H., Tan, T., *et al.* (2016). 'China's renewable energy goals by 2050', *Environmental Development*, 20, pp. 83–90.

Yaroufakis, Y. (2014). *Think Big, Think Bold: Why the Left must aim for a radical Pan-European Green New Deal*, Centre for Labour and Social Studies, London.

Zeman, F. (2007). 'Energy and material balance of CO_2 capture from ambient air', *Environ. Sci. Technol.,* 41, pp. 7558–7563.

Zencey, E. (2013). 'Energy as Master Resource', in Worldwatch Institute (eds), *State of the World 2013*, Island Press, Washington, D.C., pp. 73–83.

Zhang, J. and Smith, K. R. (2003). 'Indoor air pollution: a global health concern', *British Medical Bulletin,* 68(1), pp. 209–225.

Ziska, L., Crimmins, A., Auclair, A., *et al.* (2016) 'Food Safety, Nutrition, and Distribution. The Impacts of Climate Change on Human Health in the United States: A Scientific Assessment', in: Basu, R., English, P., *et al.*, (eds.), *U.S. Global Change Research Program*, Washington, DC, pp. 189–216.

Zmolek, M. A. (2014). *Rethinking the Industrial Revolution*, Haymarket Books, Chicago.

Appendix

A.1. Heat death of the universe: Is this a prescription for ultimate collapse of human civilization? (cited in Chapter 2)

The following is from the senior author:

> 'Entropy has been loosely defined as the measure of the disorder of a system; more precisely, entropy is the randomized state of energy that is unavailable to do work (Lehninger, 1965). In the classical interpretation, all processes in the universe ultimately must lead to its heat death as the potential for further change is ended. As Cardwell (1989) put it: "The cosmic role of heat, first discerned at the end of the Eighteenth Century and eloquently described by writers like Fourier and Carnot had thus, by way of Joule, Rankine and Kelvin, achieved its final definition by Clausius. This is not a balanced, symmetrical, self-perpetuating universe, as the development of rational mechanics, building on the foundations of Newton's System of the World, seemed so confidently to indicate. It is a universe tending inexorably to doom, to the atrophy of a 'heat death', in which no energy at all will be available although none will have been destroyed; and the complementary condition is that the entropy of the universe will be at its maximum" (Cardwell, 1989, p. 273)...heat death was not accepted by Engels, and most later Marxists, since this scenario embodies a deeply pessimistic perspective of natural evolution. In Engels' Dialectics of Nature (1987), he asserts that the heat radiated into space must by some as yet unknown mechanism be re-utilized in the eternal cycle since motion in the universe is inexhaustible

(see pp. 561–563 and p. 334). Haeckel (1900) shared Engels' view of the inexhaustibility of motion in the universe while accepting the applicability of the Second Law in local systems (pp. 246–247)' (Schwartzman, 1996, pp. 308–309).

Foster and Burkett (2008) make the case that Engels' views were the same as Haeckel's, but we find their argument not very persuasive given the quotation from Engels' *Dialectics of Nature*. Foster and Burkett (2008) discuss the Second Law of Thermodynamics, citing Georgescu-Roegen, but fail to mention the already well-known rebuttal of his purported expansion, the so-called fourth law.

Following Engel's lead, the categorical rejection of heat death became accepted canon in official Marxism–Leninism:

> 'The theory of the heat death of the universe is completely unfounded and ignores the law of conservation [sic] and transformation of energy which asserts the indestructibility of motion not only quantitatively but also qualitatively, i.e., that motion cannot exist in only one form' (Afanasyev, n.d., p. 69).

Similarly, Frolov (1984) had noted:

> 'For systems consisting of an infinitely great number of particles (the Universe or the world as a whole) the concept of the most probable state loses its meaning (in infinitely large systems all states are equally probable). By taking into account the role of gravitation, cosmology arrives at the conclusion that the Entropy of the Universe grows without tending to any maximum (the state of thermal balance). Modern science proves the complete groundlessness of the conclusions of the allegedly inevitable thermal balance and thermal death of the world' (pp. 126–127).

Soviet physicists and philosophers rejected heat death from a variety of positions, including the Nobel Laureate (1962) Lev Landau, whom, as the senior author pointed out, was 'not noted for his obsequious adherence to Marxism–Leninism', who apparently rejected heat death from considerations of relativistic thermodynamics (Graham, 1987, p. 500). Likewise, the eminent physicist and Einstein scholar B. Kuznetsov could not swallow heat death as a concept:

> 'Philosophy, in particular the philosophy of Engels, and 19th Century statistical physics advanced rather convincing arguments against thermal death. Modern science, the theory of relativity and relativistic cosmology and, to no lesser extent, quantum mechanics, forces us to interpret the thermodynamics of the Universe from new standpoints that assumedly eliminate the inevitability of thermal death, although they still do not offer any concrete and unequivocal conception of the cosmic mechanism of forming temperature gradients, contrasted to thermal death' (Kuznetsov, 1977, p. 34).

In other words, we are still waiting for the mechanism Engels was convinced could turn waste heat to low entropy energy! While a near consensus of rejection was held by the materialist camp, particularly of Marxist persuasion, supporters of the heat death scenario in the 19th and 20th Century put it to good use in a broad range of ideological interventions. For example, the argument for vitalism is based on the purported anti-entropic quality of life and its evolution (e.g., Henry Adams, following Haeckel; see Martinez-Alier, 1987). The confusion embodied in this position is easily clarified by the fact that a living organism is an open system — the entropy in the environment therefore increases as a debt for any internal process — but this erroneous position lives on in many contemporary treatments (Schwartzman, 1996, p. 310).

Contemporary cosmologists have taken a fresh look at the heat death scenario, with continued debate as to its validity in the context of cosmological theories of inflation, collapsing and expanding universes (see Davies, 1977; Barrow and Tipler, 1988; Coveney and Highfield, 1990; Barrow, 1994; Smolin, 1997, 2013; Penrose, 2011). For example, in a universe that will expand forever, consistent with the present cosmological consensus, the actual growth of entropy may never equal the maximum potential entropy, thus heat death may be indefinitely postponed (Barrow, 1994). Considering that the identity of most of the universe is still unknown, with some 95% of matter-energy consisting of dark matter/energy, it should not come as a surprise that the old heat death scenario may be reinterpreted in the future in a radically different form. For example, in some inconceivable time in the future, quantum fluctuations could produce a new big bang (Penrose, 2011). A related possibility for a no heat death end to the universe occurs with an infinite number of free energy sources (Lineweaver, 2013).

A.2. Critique of Charles Hall and others who argue that the EROEI ratios of high efficiency renewables are lower than claimed (cited in Chapter 4)

Aside from his Blog name, Tom, the author prefers to remain anonymous. This article is reproduced from http://bountifulenergy.blogspot.com/2015/05/six-errors-in-eroei-calculations.html.

'Monday, May 11, 2015

Six Errors in ERoEI calculations

Error #1: Energy returns are repeatedly treated as energy investment

As an example, the paper from C Hall (*What is the Minimum EROI that a Sustainable Society Must Have?)* calculates the EROI of oil. However, it includes the energy cost of freeways, automobiles, and so on. That is a mistake, because those things are energy returns, not energy investments to obtain energy. If I drive my car down the freeway, and I'm not doing so out of necessity for gathering coal, then it was because of energy returns.

Error #2: Lifetime estimates are incorrect

Many papers wrongly assume that the lifetime of an energy source is identical to its warranty period. For example, Hall *et al.*'s book indicated above, and Weissbach *et al.*'s paper, both assume that the lifespan of a solar PV module is 25 years because that is the warranty period.

Error #3: Not counting embedded energy which is recovered

Papers about EROI frequently include the "embedded energy" cost of components for an energy source. For example, calculations of the EROI for solar PV often include the "embedded energy" in the aluminum frames which support the solar panels in the field.

If embedded energy is counted on the way in, then it must also be counted on the way out. These papers uniformly fail to account for the energy which is recovered when the aluminum is recycled when the frames are dismantled. The recovered energy should be counted because the aluminum will be recycled. Almost all major corporations recycle structural materials such as aluminum because they save money by doing so.

This factor alone has a large effect on the reported EROI of solar PV.

Error #4: Waste heat losses are counted as energy returns

This is a recurring problem throughout the ERoEI literature. Waste heat should not be counted as energy returns because it is not usable as energy to society. The only exception is when the waste heat is actually used for something (such as combined heat and power plants), but this is rare.

Error #5: Outdated figures are used

Frequently, there are large discrepancies in the EROI calculations because different technologies are assumed when calculating energy inputs. For example, there are large discrepancies of the reported EROI of nuclear power. That is partly because some papers calculate the EROI using gas diffusion enrichment of uranium, while other papers calculate the EROI using centrifuge enrichment. Those different assumptions will yield very different EROIs for nuclear power, because centrifuge enrichment is so much more efficient. This factor is a large part of the energy investment for nuclear power, and so has a big effect on the resultant EROI.

When calculating the EROI of an energy source, we should use the most modern technology when calculating energy inputs. We wish to know the EROI of an energy source going forward, not the EROI of an energy source if we had built it years ago.

As an example, the paper by Weissbach *et al.*, in its calculations of the EROI of solar PV, assumes the Siemens process is used to generate solar PV grade silicon. However, that process has been supplanted by processes which use only 40 % of the energy. This factor by itself increases the EROI of solar PV in Weissbach's paper from 3.8 to 6.6.

Error #6: Invalid Comparisons Are Made

There are actually different types of EROI depending upon where the boundaries are drawn for calculations. When calculating the EROI of oil, do you include refinery losses? Energy losses for the transport of oil? Waste heat losses from the car? And so on. Each one of those calculations represents a different type of EROI. Some EROI calculations attempt to include only energy inputs used for extraction at the mine mouth, whereas other EROI calculations attempt to include every energy investment in the entire economy, such as the energy investment for building rail lines to transport the coal. Those are different types of EROI.

EROI figures should not be compared if they draw the boundaries very differently. For example, there was a very famous graph from Charles Hall which makes such comparisons (found here: [http://4.bp.blogspot.com/-kbHup-k7tzQ/TnV1pXZp97I/AAAAAAAAAD4/imhOY_5PJSM/s1600/1.jpg]. That graph spread like wildfire throughout the peak oil community. However, that graph is repeatedly comparing different types of EROI figures which are not comparable.

For example, the comparison of the EROI of coal (about 70) to nuclear (about 10), taken from that graph. There is a big difference in the kinds of EROI for those two sources. The figure for coal is before waste heat losses are subtracted for generating electricity, whereas the figure for nuclear is after waste heat losses are subtracted. When a correction is made for that, coal has an EROI of about 24.5, compared to nuclear of 10. The discrepancy has been reduced considerably.

As another example from the same graph, oil from 1930 is reported to have an EROI of 100, whereas hydroelectric is reported to have an EROI of 30. However, the EROI of oil from does not include refinery losses and waste heat losses from internal combustion engines in 1930. Correcting these factors yields an EROI of 10.5 for oil in 1930, not 100. Of course, hydroelectric also suffers from electrical resistance losses which reduces its EROI to perhaps 25. However, the adjusted EROI ratio for oil has gone from much higher to much lower when an adjustment is made so the figures are comparable.

Conclusion

The six errors described above are widespread throughout the ERoEI literature. They are partly responsible for the wide discrepancy between reported ERoEI findings.

For example, Charles Hall *et al.*'s book, *Spain's Photovoltaic Revolution*, is committing errors #1, #2, #3, and #5. When I correct those errors and recalculate, I obtain an EROI of 6.27 for solar PV, not 2.79 as reported.

Weissbach's paper calculates an EROI of solar PV at 3.8. However, that paper is committing errors #2, #3, and #5. When I correct those errors, I obtain an EROI of 12.96, and not the 3.8 which that paper reported. Incidentally, that paper also calculates the EROI for solar in a cloudy site in Germany, and then generalizes that to the EROI of "solar PV" altogether. If I correct that factor also, and use the average insolation for the inhabited northern hemisphere, then I obtain an EROI figure of 22 for solar PV, which is much higher than the reported figure of 3.8.

Finally, even the concept of EROI has problems. Perhaps net energy should be expressed or reported differently, using a different ratio. This is because EROI gives an exaggerated impression of the difference between energy sources. For example, assume a hypothetical energy source with an EROI of 10,000, and compare it to an energy source with an EROI of 10. The source with an EROI of 10,000 would require 0.0001% of its output (1/10000) to build another like it, whereas the source with an EROI of 10 would require only 10% of its output (1/10) to build another like it. In other words, a reduction in EROI of 99.99% led to a reduction of net energy output of only 10%. This is because EROI is less and less important as it becomes higher. Instead of using EROI, we should calculate net energy as 1-(EI/ER), and then express that as a percentage. For example, if natural gas has an EROI of 15 (everything included such as infrastructure), and solar PV has an EROI of 6.27 (everything included), then their inverted ratios are 93% and 84% respectively. This means that 7% of the energy from the gas plant is necessary to build another gas plant, whereas 16% of the energy from the solar plant is necessary to build another solar plant. The net energy available to society has declined by only 9% despite EROI falling by more than half. Thus, EROI figures give an incorrect impression, and should be calculated and reported differently.

When all the problems above are corrected, it's unclear if there is any significant difference in net energy between different methods of generating electricity. The highest EROI source (hydroelectric) requires 1.3% of its output to build another hydroelectric dam, whereas the lowest source (solar PV) requires 16%. This implies only that we would need to build slightly more solar cells (about 15% more) to obtain the same net energy. Any EROI more than 5 or so makes little difference (20% at most). All common methods of generating electricity seem to exceed that threshold.

Certainly, we should investigate further into this matter. If any method of generating electricity has a disastrously low EROI (lower than 4 or so, everything included) then it would be very helpful for us to know about it. Hall's work is very useful in this regard, insofar as he attempts to include all energy investments, which will give us better approximations of relative EROIs. However, we must avoid the above-mentioned errors in performing our calculations.'

The author, Tom, also discusses the myth surrounding ERoEI ratios of renewables in comparison to fossil fuels. The following is reproduced from http://bountifulenergy.blogspot.com/2014/07/renewables-have-higher-eroei-than.html.

'Sunday, July 27, 2014

Renewables have higher ERoEI than fossil fuels

One the central claims of the peak oil/energy decline movement, is that renewable sources of power have extremely low ERoEI. Therefore, it is claimed, renewables are no substitute for fossil fuels, because they cannot provide enough "net energy" to power civilization. In support of this claim, energy decline adherents often post graphs like this one: [http://www.theoildrum.com/files/CH6.png] showing that renewables (especially solar PV) have low ERoEI compared to fossil fuels. More recently, Hall and Prieto have published a book, Spain's photovoltaic revolution, in which they claim that the ERoEI of solar PV in Spain is only 2.45, which is far lower than the ERoEI of fossil fuels.

In fact, those claims are entirely wrong. Renewables have ERoEI ratios which are generally comparable to, or higher than, fossil fuels.

Although peak oilers reach a different conclusion, that is because they are carrying out the calculation incorrectly. They are ignoring or not including massive waste heat losses (generally 60% or more) from combustion engines which drastically reduces the ERoEI of fossil fuels. Those waste heat losses provide no energy services to society, and should be counted as losses, but are wrongly counted as "energy returns" by peak oilers. Furthermore, peak oilers are ignoring or not counting other large energy losses of fossil fuels. Those omissions exaggerate the ERoEI of fossil fuels relative to renewables. When the calculation is carried out correctly, renewables have higher ERoEI ratios than fossil fuels.

In other words, the notion that renewables have ERoEI ratios which are lower than fossil fuels, is simply mistaken. It arises from performing invalid, apples-to-oranges comparisons, or from not counting energy losses of fossil fuels.'

References cited in this Blog

Hall, C. A. S., Balogh, S. and Murphy, D. J. R. (2009). 'What is the Minimum EROI that a Sustainable Society Must Have?', *Energies*, 2, pp. 25-47.

Prieto, P. A. and Hall, C. A. S. (2013). *Spain's Photovoltaic Revolution: The Energy Return on Investment,* Springer, Berlin.

Weissbach, D., Ruprecht, G., Huke, A., *et al.* (2013). 'Energy Intensities, EROIs (Energy Returned on Invested), and Energy Payback Times of Electricity Generating Power Plants', *Energy* 52, pp. 210–221.

Index